Inquiry-Based Science Education

T0331550

Global Science Education

Professor Ali Eftekhari
Series Editor

Learning about the scientific education systems in the global context is of utmost importance now for two reasons. Firstly, the academic community is now international. It is no longer limited to top universities, as the mobility of staff and students is very common even in remote places. Secondly, education systems need to continually evolve in order to cope with the market demand. Contrary to the past when the pioneering countries were the most innovative ones, now emerging economies are more eager to push the boundaries of innovative education. Here, an overall picture of the whole field is provided. Moreover, the entire collection is indeed an encyclopaedia of science education and can be used as a resource for global education.

Series List:

The Whys of a Scientific Life
John R. Helliwell

Advancing Professional Development through CPE in Public Health
Ira Nurmala and Yashwant Pathak

A Spotlight on the History of Ancient Egyptian Medicine
Ibrahim M. Eltorai

Scientific Misconduct Training Workbook
John Gaetano D'Angelo

The Whats of a Scientific Life
John R. Helliwell

Inquiry-Based Science Education
Robyn M. Gillies

Hark, Hark! Hear the Story of a Science Educator
Jazlin Ebenzer

Inquiry-Based Science Education

Robyn M. Gillies

CRC Press
Taylor & Francis Group
Boca Raton London New York

CRC Press is an imprint of the
Taylor & Francis Group, an **informa** business

CRC Press
Taylor & Francis Group
6000 Broken Sound Parkway NW, Suite 300
Boca Raton, FL 33487-2742

**Visit the Taylor & Francis Web site at
http://www.taylorandfrancis.com**

**and the CRC Press Web site at
http://www.crcpress.com**

Contents

1 Inquiry-Based Science   **1**
Introduction   1
Background   1
Inquiry-Based Science   2
Using Inquiry-Based Science to Challenge Thinking   5
Strategies Promoting Inquiry-Based Science   11
Challenges Implementing Inquiry-Based Science   18
Chapter Summary   19
Additional Readings   20

**2 Visual, Embodied, and Language Representations in
Teaching Inquiry-Based Science: A Case Study**   **21**
Introduction   21
Types of Representations   22
Method   24
Results and Discussion   26
Chapter Summary   42
Additional Readings   42

3 Developing Scientific Literacy   **43**
Introduction   43
Background   43
Scientific Literacy   44
Chapter Summary   61
Additional Readings   62

4 Promoting Scientific Discourse   **63**
Introduction   63
Dialogic Teaching   64
Strategies to Promote Dialogic Interactions   70
Dialogic Strategies for Students   76
Chapter Summary   79
Additional Readings   79

5 Structuring Cooperative Learning to Promote Social and Academic Learning   **81**
   Introduction   81
   Cooperative Learning   82
   Benefits of Cooperative Learning   83
   Key Elements in Cooperative Learning   86
   Strategies for Constructing Cooperation in Groups   93
   Strategies for Assessing Cooperative Learning   94
   Chapter Summary   96
   Additional Reading   97

6 The Structure of Observed Learning Outcomes (SOLO) Taxonomy: Assessing Students' Reasoning, Problem-Solving, and Learning   **99**
   Introduction   99
   The SOLO Taxonomy   100
   Five Levels of the SOLO Taxonomy   103
   Intended Learning Outcomes   104
   Chapter Summary   107
   Additional Readings   108

References   109
Index   113

Inquiry-Based Science

1

INTRODUCTION

This chapter provides an introduction to inquiry teaching in science, models for teaching inquiry, and approaches to evaluating the inquiry process. In recent years, emphasis has been on teaching science using an inquiry approach where students are actively involved in scientific investigations that challenge their curiosity, encourage them to ask questions, explore possible solutions to problems, use evidence to explain phenomena, elaborate on possible effects, evaluate findings, and predict potential outcomes if different variables are changed. This chapter also presents examples of how students are cognitively challenged to make sense of the phenomena under investigation, develop evidence-based explanations, and communicate their ideas and understandings in discipline-specific language as to why solutions to problems work and others do not.

BACKGROUND

Over the last two decades, emphasis has been on teaching science through inquiry. Inquiry-based science adopts an investigative approach to teaching and learning where students are provided with opportunities to scrutinise a problem, search for possible solutions, make observations, ask questions, test out ideas, and think creatively, and in so doing, learn to reconcile their developing understandings with previous knowledge and experience. Inquiry has many potential benefits. When students are involved in inquiry-based science, they are *doing science* where they are learning the processes communities of scientists employ to investigate phenomena. In so doing, they learn to explore possible solutions, develop explanations for the topic under investigation, elaborate on concepts and processes, and evaluate or assess their understandings in the light of the evidence available to them. This approach to teaching relies on teachers recognising the importance of presenting problems to students that will challenge their current conceptual understandings so they are forced to reconcile anomalous thinking and construct new conceptual understandings.

Cultivating students' scientific habits of mind, developing their capabilities to engage in scientific inquiry, and teaching them how to reason in the scientific context is one of the principal goals of science education (National Research Council, 2012, p. 41). In fact, the essential elements in any science education programme must include: (a) the development of conceptual understanding; (b) the improvement of cognitive reasoning; (c) the improvement of students' understanding of the epistemic nature of science; and (d) the affordance of effective experiences that are both positive and engaging (Osborne, 2006). Furthermore, this needs to occur within the context of social practices and values that both promote and sustain the scientific enterprise and lead to the production of reliable knowledge.

When students have opportunities to engage with their peers in collaborative scientific inquiries, they learn to ask questions about different phenomena, plan investigations, use a variety of tools and artefacts to collect and analyse data, and use evidence to develop claims and propose possible explanations for the phenomena they have observed (Bell et al., 2010; Llewellyn, 2014). In inquiry-based science, students not only learn the relevant content but also learn the discipline-specific reasoning skills and practices by collaboratively engaging in authentic problems or questions with their peers. In so doing, students are cognitively challenged to make sense of the phenomena under investigation, develop explanations that are based on evidence, and communicate their findings in discipline-specific language as to why certain solutions to a problem work and others do not.

When you have finished this chapter, you will know:

- What inquiry-based science is.
- How inquiry-based science challenges students' thinking.
- Strategies teachers can use to promote inquiry-based science in their classrooms.
- Challenges teachers face when implementing inquiry-based science in their classrooms.

INQUIRY-BASED SCIENCE

Inquiry-based science is an investigative approach to teaching and learning where students are provided with opportunities to investigate a problem, search for possible solutions, make observations, ask questions, test out ideas, think creatively, and use their intuition. The inquiry process is complex as it involves students reconciling their current understandings with both the evidence obtained from an inquiry and the ability to communicate their newly acquired knowledge in a way that will be accepted as well-reasoned and logical. Such a process is challenging, requiring teachers to play an active role in helping students learn the steps in the inquiry process.

Scientific inquiry recognises the diverse ways in which scientists study the natural world and propose explanations based on the evidence derived from their work. It also refers to "the activities through which students develop knowledge and understanding of scientific ideas, as well as an understanding of how scientists study the natural world" (National Science Teachers Association, 2004, p. 1). When students have opportunities to engage in scientific inquiry, they learn to use their ideas and, in so doing, deepen their conceptual understanding of scientific content as well as their understanding of how to do science. "This science-as-practice perspective brings together content knowledge and process skills in a manner that highlights their interconnected nature" (Harris & Rooks, 2010, p. 229), facilitating student engagement with complex science ideas and participation in scientific activities. In effect, students gradually learn to understand the practices that scientists engage in when confronting various scientific problems (Herrenkohl et al., 2011).

Inquiry is the process of investigating a problem issue that requires critical thinking, observing, asking questions, testing out ideas and hypotheses, and engaging in collaborative discussions to communicate scientific knowledge and develop explanations or solutions on the topic under discussion (Lee et al., 2004; Metz, 2008). While children often demonstrate a natural curiosity about the world in which they live, research indicates that they rarely ask questions about what they have seen and heard. Helping students to understand the inquiry process where they learn to ask questions about phenomena that challenge their current understandings, propose possible explanations for what they see, and reconcile understandings with their current knowledge to create new knowledge and understandings takes a concerted effort on the part of the teacher. While there are many approaches to teaching students how to engage in inquiry, Figure 1.1 represents generally agreed steps in the process.

Inquiry learning is seen as critically important to helping students engage in science, yet teachers continue to struggle with what inquiry should look like and how it should be taught. Zuckerman et al. (1998) identified three factors that they considered crucial for teaching inquiry science to primary and middle years students. These factors are

1. Arousing students' imagination by presenting new and awe-inspiring phenomena that are already within students' current level of development so the child has the capacity to recognise the new

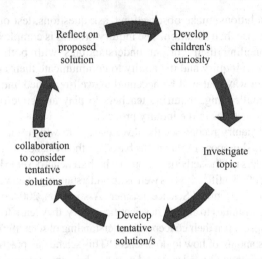

FIGURE 1.1 Steps in the inquiry process.

elements in the phenomena and to connect these new elements to the context and structure of existing background knowledge and experience.

2. Teachers need to provide opportunities for students to work with others to investigate, discuss, and resolve challenging problems.
3. Students need to be encouraged to participate in asking questions to help them test out their ideas and eventually verify their hypotheses.

The promotion of inquiry is highly dependent on the teachers' efforts to guide and scaffold students' learning as they engage in the inquiry process, so they understand how to think as they participate in tasks, as well as acquire the procedural knowledge of how to complete these tasks (Duschl & Duncan, 2009; Veermans et al., 2005). This involves challenging children's thinking and problem-solving by making explicit the types of thinking they need to demonstrate. When this occurs, Gillies and Boyle (2006) found that children, in turn, are more focused and explicit in the types of responses they provide and the help they give to each other.

Given that inquiry usually involves collaborative discussions, students need to know how to cooperate with peers so they listen to what others have to say, share ideas and information, clarify misconceptions, generate new understandings, and critically reflect on what they have learned and what they still need to learn. In fact, when this happens in science classrooms, Ford and Forman (2015) argue that students engage in a process of dialogical discourse that encourages them to collaboratively construct and critique different ideas and points of view and, in so doing, they begin to learn how to function as a scientific community. This give and take in discussions, Ford and Forman believe, is essential if productive scientific talk is to occur. Moreover, it is this dialogical discourse that, in turn, supports changes to students' reasoning and scientific habits of mind or way of reasoning that promotes problem-solving, insightfulness, perseverance, creativity, and craftmanship (Costa & Kallick, 2000).

USING INQUIRY-BASED SCIENCE TO CHALLENGE THINKING

Inquiry-based science challenges students' thinking by engaging them in investigating scientifically orientated questions where they learn to prioritise evidence, evaluate explanations in the light of alternative explanations, and communicate and justify their decisions in language that is specific to science. However, in a review of 225 studies between 1972 and 2011, Howe and Abedin (2013) found that classroom dialogue is dominated by teacher–student initiation-response feedback (I-R-F) (e.g., Teacher: Who was the first man on the moon? Student: Neil Armstrong. T: Yes, that's right), which tends to only require minimal responses with no elaboration. Unfortunately, research indicates that students rarely engage in classroom-based discourse where they ask question, discuss issues, or provide reasons for the positions they have taken. On the other hand, Mercer and Sams (2006) found that when students were taught how to use language as a tool for thinking and reasoning, they were able to use talk to think and reason more effectively. In a similar vein, Gillies and Baffour (2017) found that when teachers spent time interrogating students' understandings and scaffolding and challenging their thinking, the students, in turn, were more attentive and used more sophisticated scientific language to explain the phenomena they were investigating than students in classrooms where teachers did not emphasise these practices.

There is no doubt that teachers play a key role in inducting students into ways of thinking and reasoning by making explicit how to express ideas, seek help, challenge different propositions, and reason in a well-argued and cogent manner. While research clearly indicates that when teachers make use of these dialogic strategies, students' participation in class and their educational achievements are likely to benefit (Mercer & Dawes, 2014), many teachers are still reluctant to embrace these strategies, preferring to utilise a transmission model of teaching where the teacher controls the channels of communication and the students remain as passive recipients.

One instructional approach that has been used successfully to teach inquiry science that challenges children's thinking and learning is the 5Es Instructional Model (Bybee, 2014). This model of teaching is research based and highlights the importance of cooperative learning where students work together in small groups to resolve problems. It also recognises the importance of students engaging in activities that challenge their current conceptions (or misconceptions) with opportunities provided to enable them to restructure their ideas and abilities.

The 5Es Instructional Model consists of five phases that Bybee (2014) believes is iterative with teachers recycling through this approach as needed. The five phases are Engagement, Exploration, Explanation, Elaboration, and Evaluation.

1. **Engagement.** The goal of this phase is to capture the students' attention and curiosity through the presentation of a novel event, situation, demonstration, or problem that involves the content and the abilities the lesson is designed to teach. For example, if students were about to embark on learning about earthquakes, the presentation of video on a tsunami and the population affected would be an example of an activity designed to engage students' attention and curiosity. Follow-up questioning by the teacher will help to challenge students' thinking as they consider the implications of such an event. For example,
   * *What do you think may be the impact of this event on people's lives?*
   * What sort of planning do you think people may need to do if they live in areas that are prone to earthquakes?

The purpose of this phase is to attract students' attention and interest in the topic with the intention of motivating them to explore or investigate the topic in more depth. Activities associated with this phase may include developing a Think, Want, Learnt, How (TWLH) chart where the students identify what they currently know about earthquakes, what they want to learn, what they have learned, and how they know.

TWLH Chart

WHAT WE THINK WE KNOW	WHAT WE WANT TO LEARN	WHAT WE LEARNED	HOW WE KNOW

The TWLH chart is used to assess students' current understandings and beliefs about the topic with the intention of helping them to identify what they still want to learn. The process is very much a guided inquiry as the teacher probes the students' knowledge and understanding and gauges their abilities to reconcile new and challenging information into their cognitive schema. This phase also provides opportunities for the teacher to informally uncover any misconceptions that students have in order to plan activities and experiences to help students explore the topic in more depth.

2. **Exploration.** This next phase focuses on providing opportunities for students to explore the topic in more depth. This may include through electronic searches, group discussion, a visit from a scientist who can elaborate on the topic, or a field trip to gather information. Consequently, students would be expected to be able to describe the difference between terms associated with earthquakes, discuss the use of different scales for measuring earthquakes, and analyse different numerical and factual information. These activities would occur in the context of group discussions where students share their information and findings, read and analyse factual information together, and identify questions that need to be resolved.

Questions such as the following may be posed by the teacher to help students explore the topic in more depth:

- *What happens when an earthquake occurs?* Describe what you have learned from your exploration of this topic.
- What instruments are used to measure the strength of an earthquake?
- What is the difference between the Richter and the Modified Mercalli scales? Describe the advantages and disadvantages of each.
- What happens to the tectonic plates when they are subject to different stresses? Describe the effects.

3. **Explanation.** The scientific explanation for the phenomena under investigation is actively pursued during this phase with the teacher directing students' attention to key parts of the previous phase while "pressing" students for their explanations. Building on students' explanations and experiences, the teacher introduces key concepts and technological terms, including the relevant scientific vocabulary and practices that help to make the explanations clear. It is important that the students are introduced to activities that are challenging yet achievable with scaffolding by the teacher if needed. Activities where students learn to construct multimodal explanations drawing

on a range of representations (e.g., tables, pictures, oral presenta-
tions, videos, and models) are undertaken during this phase. Specific
examples may include:

(a) Using written language and models to demonstrate their
    understanding of earthquakes and tectonic plates.
(b) Using scientific language to describe three types of tectonic
    plate movements and their effects on the earth's crust.
(c) Constructing a portfolio on a topic that is designed to provide
    an ongoing record of work that students have attempted or
    have completed. Portfolios provide insights into students'
    abilities to communicate scientifically, demonstrate sci-
    entific reasoning, and make connections between differ-
    ent concepts and relationships. They also enable students
    to reflect on the progress they have made and what they
    still need to do if they wish to achieve. This activity can
    be conducted in conjunction with the class teacher when
    students discuss personal learning goals or as part of a
    group activity where the group identify what they want to
    achieve.

During this phase, it is critically important that the teacher asks thought-provoking questions to help students think deeply about the topic they are investigating. The following are examples of such questions:

- *Explain why or how….?*
- What is the difference between … and …?
- *What do you think could happen if …?*
- What do you think causes … and why?
- What is the evidence that supports this statement?

4. **Elaboration.** This phase builds on the previous phase so students
   are encouraged to elaborate on their conceptions using additional
   information and understandings. During this phase, the teacher
   actively challenges students' current conceptions and skills by
   providing additional experiences that will help them to develop
   new insights and broader understandings of the topic. For example, students may be discussing how movement of the earth's tectonic plates can create earthquakes that can occur on land or in
   water. The teacher may build on these understandings by challenging the students to elaborate on how tectonic plates move

(e.g., convergent, divergent, or transform) and the different effects they generate. In so doing, the teacher encourages more in-depth analysis and elaboration on the phenomena. Students, in turn, can elaborate on their current conceptions through writing reports or producing portfolios, participating in debates that challenge current conventions, or utilising diagrammatic and graphic modes to present information that provides additional insights on the topic at hand.

Additional activities may include:

(a) Constructing a seismometer to illustrate how data on seismic waves are collected. Students work in small groups to construct the seismometer and demonstrate how data can be collected from it.

(b) Interview a seismologist to determine what this scientist does, how information is collected and interpreted, and how that information is communicated to the wider community.

(c) Work in small groups to build models to withstand weak and strong simulated earthquake movements and elaborate on the advantage and disadvantage of each. Attention should be directed at ease of construction, cost of materials, aesthetic appeal, and impact on the population affected.

Questions that could be used to challenge and scaffold students' elaborations include:

- *Perhaps you can provide further information on how and when seismic data are collected by seismologists and what they do with these data?*

- Many people in the population would find it difficult to interpret seismograms so I wonder if there may be other ways in which this information can be communicated?

- Perhaps you can elaborate further on how seismograms can be used to help people understand the consequences of living in earthquake-prone regions?

5. **Evaluation.** This final phase provides teachers and students with the opportunity to review the progress the students have made in developing different scientific understandings through both informal and formal assessments. Informal assessments can include the collection

of various artefacts (e.g., journals, portfolios, models, exhibitions of performance) that demonstrate different conceptual understandings, while formal assessments may include responses to specific tests designed to ascertain students' conceptual understandings of the topic.

During this phase, teachers need to provide opportunities for students to reflect on their progress. This may be done in a one-on-one conference where the teacher interviews each student to ascertain what they have learned and what they may still be struggling to understand. The language students use during this case conference is just one way of gauging how the students are using different scientific terms and language in response to questions asked.

Another approach to encouraging students to reflect on their learning involves using the following Know, Learned, and Questions raised (KLQ) chart. This chart acts as an organiser to help students discuss their responses to these probes. This activity can be undertaken individually or as part of a small-group activity. The advantage of this type of activity is that the chart provides a structure that enables teachers to promote thinking, reflection, and metacognitive processes in a coherent fashion by asking students to recall what they know and have learned as well as think metacognitively by reflecting on what questions remain unanswered. These are thinking processes that successful learners demonstrate.

KNOW	LEARNED	QUESTIONS RAISED

Questions that can be asked during this phase may include the following five types of questions that King (1997) identified as part of a sequence of questions to promote higher-level thinking:

- "Describe ... in your own words" (Review questions)
- "Tell me more about ..." (Probing questions)
- "Have you thought about ...?" (Hint questions)
- "What is the difference between ... and ...?" (Intelligent-thinking questions)
- "Have I covered all the points I need to?" (Self-monitoring questions)

STRATEGIES PROMOTING INQUIRY-BASED SCIENCE

"Scientific inquiry requires the use of evidence, logic, and imagination in developing explanations about the natural world" (Newman et al., 2004, p. 258). In inquiry-based science, students work together in cooperative small groups to investigate topics, share information that they have found, and discuss and evaluate different explanations that may explain the phenomena. This process is iterative until they can communicate and justify their explanations in the context of the investigation they are undertaking.

Cooperative Learning Activities

Successful cooperative learning activities involve students working together, listening to each other's ideas, trying to understand different perspectives, suggesting alternative explanations for the phenomena, and working together constructively to accept responsibility for completing their part of the task while assisting others to do likewise. When this happens, Ford and Forman (2015) argue that students engage in a process of dialogical discourse that encourages them to cooperatively construct and critique different ideas and perspectives and, in so doing, they begin to learn how to function as a scientific community. Ford and Forman maintain that this type of interaction is essential if productive scientific talk is to occur. Moreover, it is this dialogical discourse that, in turn, supports changes to students' conceptual understandings and reasoning and scientific habits of mind.

Strategies to help students learn to work cooperatively together include:

1. **Brainstorm with a Peer.** Have students work with the student beside them to brainstorm some ideas from the lesson. Jot down six ideas. Allow 2 minutes for this activity. The teacher then calls on different dyads to report what they discussed. The advantage of this type of activity is that it helps students to learn to listen to others and consider their ideas.
2. **Paired Activity.** Students interview each other about their favourite DVD, sport, activity, book, and so on. The students spend 2 minutes on this activity. The teacher then calls on specific dyads to introduce each student to the class. As a follow up to this activity, it is important for the teacher to discuss with the class whether the students now have a better understanding of the person who was being

introduced and what questions might need to be asked to provide clearer information. The advantage of this activity is that it makes students aware of other students' interests and, because they will be required to introduce the other student to others, they have to actively listen to what is said.

3. **Listen and Recall.** Students work in pairs on a topic and jot down the main ideas. One adopts the role of the listener while the other recalls the information they have learned. The listener tries to ask questions to help clarify issues or assist the other recall what was learned. Questions such as the following are used to probe and clarify issues:

   - What do you mean by …?
   - Can you tell me more about …?
   - What would happen if …?

After 5 minutes, the students change roles and the process of interrogating the topic begins again. As students learn to ask more questions to help clarify the issues they are discussing, their questions become more detailed and the responses more elaborated. Moreover, King (1999) found that by encouraging the listener to ask more thinking questions, the recaller is more likely to respond with explanations and elaborations or the types of responses that are known to promote learning. Eventually, as the students learn to think more deeply about the information they are discussing, they learn to ask more metacognitive questions or questions that demonstrate how they are thinking about the topic.

4. **2-Minute Review.** The teacher stops at any time during the lesson and gives students (working in pairs) 2 minutes to recall aspects of the lesson. Students are then called on by the teacher to discuss what their dyad identified. The advantage of this review is that once students get used to this routine, it helps them to stay "tuned in" to what the lesson is about. As most children will not readily be able to recall all aspects of the lesson, they will rely on their peer to assist with this task, thereby demonstrating interdependence with "two heads better than one"; a key element of successful cooperative learning (Gillies, 2007).

5. **Paired-Questioning.** Students read a passage together and then ask each other a set of specific questions to help clarify their understanding of it. It may be necessary to cue students' questioning by giving them a set of question stems to guide their questioning. For example:

- What is the main idea of ...?
- Explain why ...?
- Explain how ...?
- How are ... and ... similar?
- What is the difference between ... and ...?
- How does this relate to what I've learned before?
- What did you like about ...?

The advantage of this activity is that students learn how to ask progressively more difficult questions as they seek to clarify their understandings of the information.

6. **Think-Pair-Share.**
   Students work in pairs on a topic. Pairs then join another pair to form a group of four. One pair shares the information and ideas they have with the other pair, then the other pair shares their information and ideas. Students are then required to develop a common list of points or ideas. Students number themselves from 1 to 4 as the teacher asks a number (i.e., student) from each group to discuss an idea their group identified as important and why they chose this idea.

There are two advantages to this approach:
1. Students need to listen to what the group members have been discussing if they are to present an idea the group have discussed.
2. The student who responds presents an idea the group have discussed rather than an individual's idea. This helps to reduce anxiety during the feedback session.

Other strategies that assist cooperation include:

Group size
Students work best in groups of two, three, or four members, simply because it is easy to hear and see what the group is doing. In larger groups, it is easier for students to passively participate as others may dominate the discussion, the roles, and the resources with little regard for less active students.

Group composition
The composition of the group is also important as research indicates that students generally work better when

- Groups are mixed in ability (high, medium, and low), although teachers need to be careful to ensure that low-ability students are not too overwhelmed by the group.
- Mixed gender.
- Status is provided to low-status students with an emphasis on the strengths a particular student brings to a group.

Type of task
There are a range of tasks that students can undertake to help them learn how to work cooperatively together. These include simple and complex tasks.

Simple tasks involve:

- Brainstorming ideas
- Recalling basic information
- Jotting down main ideas on a topic

Complex tasks will require the students to problem-solve together. This may involve:

- Identifying possible solutions to a problem and justifying answers.
- Identifying possible solutions to a problem, including both the positive and negative consequences, choosing the best solution, justifying the answer, and then developing a logo, text message, or advertisement that clarifies this choice and justification.
- Identifying a list of questions that could be asked to help clarify the problem.

Complex tasks that challenge thinking are constructed so there is no right answer, requiring students to discuss how to proceed. This type of task is usually completed in small groups where students are expected to work together to contribute ideas, discuss the perspectives and ideas of others, and evaluate possible solutions in the light of the information presented.

Students engaged in challenging tasks are also encouraged to evaluate the process the group employed in working towards a solution and the outcomes achieved. This can be achieved by asking students to reflect on:

what we have achieved;
what we still need to achieve; and
how might we do this.

Individual Reflection Activity

SKILLS	THIS IS HOW I RATE MYSELF		
I used positive statements to encourage my group members.	⬆ Always	⬆ Sometimes	⬆ Never
I contributed my ideas and information.	⬆ Always	⬆ Sometimes	⬆ Never
I asked others for their ideas and information.	⬆ Always	⬆ Sometimes	⬆ Never
I helped others in my group learn.	⬆ Always	⬆ Sometimes	⬆ Never
I helped the group organise and write up the group's ideas and information.	⬆ Always	⬆ Sometimes	⬆ Never
I stayed on task and followed my group role.	⬆ Always	⬆ Sometimes	⬆ Never
I included everyone in our work.	⬆ Always	⬆ Sometimes	⬆ Never

Group evaluation: The following rubric can be used to help students evaluate the progress of their group:

Group's Action Plan

Group's goal is: ....

TASKS	WHO DOES WHAT	EVALUATION		
		FINISHED	NOT FINISHED	NOT ATTEMPTED

Overall comments on the group's progress:................................

Characteristics of Complex Tasks

- Multiple roles for participants based on learner strengths (desktop publisher, media manger, production manager, personnel manager).
- Multiple subtasks that contribute to the larger group task with each group member contributing.
- Discussion is necessary so students understand that it is acceptable to talk and seek and give help to other group members. Students learn to ask for help and keep asking for help until it is given and that it is important to provide explanations and not just minimal responses.
- Group product is the expected outcome. This may include a PowerPoint, diorama, information chart, role play, performance, or portfolios that illustrate the learning that has occurred.
- Students are taught to reflect on the process and outcomes. (What did we do that worked well? What do we still need to do? How can we do it?)
- Criteria for task completion are clearly stated and checked off on the criteria sheet.

Ways to Evaluate Students' Learning from Working on Complex Tasks

- Quality of the discussion can be determined by the questions asked and responses given (higher-level thinking questions that elicit explanations), depth of discussion (conceptual understandings expressed), and justifications and reasons provided.

- Product outcome – comprehensive, covered key facets of the problem, creativity in response. For example, the word web shown in Figure 1.2 may be one way of evaluating how students are linking key concepts to the earthquake topic.
- Process employed – inclusive of others, respectful to others, willingness to consider others' points of view.
- Student reflections on the activity – what they perceived they learned from it. For example, provide two or three questions to promote the thinking about what has been discussed or experimented with during the course of the lesson. Present each question one at a time:

What was the most interesting thing that you learned today?
What would you like to learn more about?
Write a question about an idea or experiment that could help your group to think about one of the issues in the lesson.

Note: The completed team word web provides a natural tool for assessing group functioning; if each student writes in a different colour and the colour code is placed at the bottom of the team word web, the teacher can see the contributions made by each team member. It can also be a very interactive activity that generates a lot of focused discussion among students (Figure 1.2).

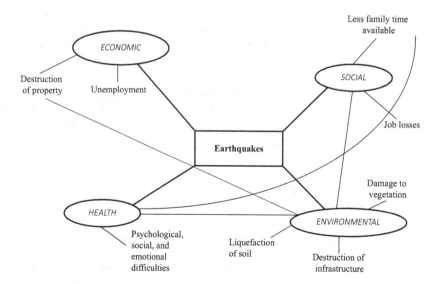

FIGURE 1.2 Team word web on earthquakes.

CHALLENGES IMPLEMENTING INQUIRY-BASED SCIENCE

One of the challenges teachers confront in teaching inquiry-based science is the misconception that they hold about what inquiry science involves. Many teachers, for example, often think they are "doing inquiry" because they are out at the front of the classroom directing the inquiry or demonstrating how to do it. This is not inquiry science. Inquiry science requires teachers to be able to excite the students' interest in a topic and then provide them with opportunities to undertake the investigation either by themselves or preferably in collaboration with others. The teacher, though, needs to remain active in the lesson, guiding the students, asking questions to help them consolidate their understandings, and providing feedback when needed to help students reflect on how they are progressing.

When students have opportunities to engage in scientific inquiries, they learn to use their ideas and, in so doing, deepen their conceptual knowledge and understanding of scientific content as well as their understanding of how to engage in doing science. Opportunities to experience science by doing it helps them to reconcile content knowledge with process skills, enabling students to engage more successfully with complex science ideas. Teachers can gauge the success of their teaching through students' level of engagement with the topic and each other, the scientific language students use to communicate their ideas, and the quality of the work they produce. Subtle comments such as "Are we doing science today? I really liked the way we did..." are typical of the types of comments students will make when they enjoy participating in science investigations.

A second challenge teachers confront in teaching inquiry-based science is how to establish small groups so students have opportunities to collaborate on topics that they are investigating. Placing students in ad hoc groups and expecting them to cooperate does not always guarantee that they will. Research demonstrates that groups are more likely to cooperate when they are well-structured so students understand how they are to work together, contribute information and ideas, accept responsibility for completing the tasks assigned to them, and assist others' to do likewise. When groups are established so these elements are evident, they are referred to as well-structured groups. In contrast, groups that are unstructured have many of the characteristics of traditional, whole-class settings where there is no requirement for students to work together to achieve the group's goal, leaving students to either work in competition with each other or individually to achieve their own ends.

Structuring a Cooperative Learning Activity

Students are organised into groups of four members and provided with the following visual organiser. Each student collects information on the topic and inserts that information in one of the quadrants to share with others in their group. This activity provides students with the opportunity to record what they know about a topic and then to negotiate with group members to select the best ideas to be inserted in the oval in the centre of the organiser. For example, students may be asked to respond to the following questions: "What are some various kinds of micro-organisms? Why is it important to know about them?" Once students have agreed on the best ideas, one member of the group then reports on these ideas to the larger class where other students, in turn, have opportunities to question the group about their selected ideas.

Visual Organiser for Cooperative Group Work

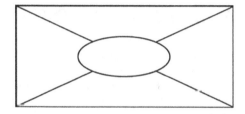

CHAPTER SUMMARY

This chapter has highlighted the importance of engaging students' interest in inquiry science by having them do science where they learn to investigate topics together and engage in processes communities of scientists employ when seeking solutions to problems at hand. In doing science, students learn to explore possible solutions, develop explanations for the topic under investigation, elaborate on understandings, and evaluate or assess their conceptions using discipline-specific reasoning skills and practices. There is no doubt that good teachers engage students' interest through novelty, something unusual that spurs their curiosity, and then use language that is very dialogic or language that lets students know that they are interested in what they think or want to say about the topic. Good teachers, then, carefully guide students as they begin to explore or investigate the topic, being careful not to dominate the conversation but allow students time to develop responses or think about the issue more carefully. In this sense, they give students the time to reflect and think more carefully about issues. However, good teachers are always careful to ensure that the inquiry-based science lesson moves forward and they do this by asking questions that probe and challenge students' thinking as well as giving them feedback that is meaningful and timely.

Teachers who do inquiry well tend to have a very good understanding of both the content that they teach and the processes involved. They use language that is very collaborative and friendly and take a genuine interest in what students are doing. They ask questions that challenge students' thinking. There is no doubt that children will engage in higher-level thinking if teachers give them time to talk about a topic; making explicit the types of thinking they need to demonstrate. When this occurs, students tend to be more focused and explicit in the types of responses they provide and the help they give to each other; language that is associated with successful learning.

ADDITIONAL READINGS

Bybee, R. (2014). The BSCS 5 E instructional model: Personal reflections and contemporary implications. *Science and Children*, *51*(8), 10–13.

Gillies, R. & Nichols, K. (2015). How to support primary teachers' implementation of inquiry: Teachers' reflections on teaching cooperative inquiry-based science. *Research in Science Education*, *45*(2), 171–191.

Osborne, J. (2014). Teaching scientific practices: Meeting the challenge of change. *Journal of Science Teacher Education*, *25*(2), 177–196.

Visual, Embodied, and Language Representations in Teaching Inquiry-Based Science

A Case Study

2

INTRODUCTION

Teaching students to understand and use representations in science is critically important if they are to become scientifically literate and understand the different ways in which representations can be used in science (Tytler, 2007). Representations, though, are more than simply transmitting information; they are integral to reasoning about scientific phenomena (Klein & Kirkpatrick, 2010). Moreover, students need to be able to create or design new representations, appraise their suitability for different tasks, and recognise and explain the purposes of different representations. In fact, there is now a major trend towards using multiple representations that include visualisations, simulations, social interactions, and written prompts to support students' conceptual understandings, strategic and process skills, and metacognitive representations in learning science (Lin et al., 2012).

Students also represent their learning through embodied representations or representations that involve sensorimotor routines for representing different ways of thinking or cognitions (Pouw et al., 2014). When students use embodied representations such as hand movements, facial gestures, and other body actions, they develop more elaborate linguistic descriptions of the phenomena represented by these embodied representations and, in so doing, gain a deeper understanding of the topic they are investigating. There is now strong consensus that students use and interpretation of different representations in the elementary and middle years of schooling are critically important for the development of scientific literacy and conceptual understandings in science.

There is also growing recognition that students are more motivated to learn when they have opportunities to reflect on their understandings and revise their representations based on new experiences and conceptual developments (Carolan et al., 2008). In fact, it has been suggested that students should be encouraged to draw more, as drawing helps students to explore, coordinate, and justify their understandings, deepen their knowledge of how different representations are used in different contexts (e.g., line graph, table), organise information, and make their thinking explicit. Hubber et al. (2010) found that when students were encouraged to use a variety of multimodal representations, such as drawings, they not only used more scientific language to explain various concepts, but they also generated more logically coherent representations than they had previously.

When you have finished this chapter, you will know:

- How to use different visual, embodied, and language representations to engage students' interest in inquiry-based science activities.
- How to encourage students to develop scientific explanations.
- How to develop deeper conceptual understandings that transfer to novel topics.

TYPES OF REPRESENTATIONS

Representational competence is a term used to describe a set of skills and practices that enable individuals to use different representations to think, interact and communicate, explain, support, problem-solve, and predict what may happen to different phenomena. These skills are best developed and used within the context of student discourse and scientific investigations where they have opportunities to ask questions, plan investigations, analyse data, and present findings. By sharing their findings with others, students learn to exchange ideas and information, clarify any misconceptions, and provide and receive feedback from their peers. In so doing, they learn to present their findings in a way that is accepted as comprehensible and balanced.

Teachers play a crucial role in demonstrating how language can be used in dialoguing with each other to promote thinking and reasoning. They do this by modelling how to express ideas, seek assistance, contest opposing propositions, and reason cogently. Mercer and Littleton (2007) proposed that when teachers use dialogue to orchestrate and foster a community of inquiry in a classroom in which students are encouraged to take an active and reflective role in building their own understandings, students, in turn, reciprocate, building on the discourse that occurs to develop clearer understandings of the topic at hand.

Dialogic discussions are facilitated when teachers actively engage in helping students to make their thoughts, knowledge, and reasons clear; they model different ways of using language that students can appropriate, and they provide opportunities for students to engage in sustained discussions so they can express their ideas, elaborate on their own understandings, and share any difficulties they may be experiencing (Gillies, 2015). It is argued that explicitly teaching students how to think and reason in science is critically important if students are to learn that there are different ways of reasoning about the phenomena under investigation (Kind & Osborne, 2017).

Fostering student discourse is critically important if students are to learn how to share ideas, separate evidence from opinion, ask questions to clarify their thinking, and offer explanations to justify their positions (Huff & Bybee, 2013). Moreover, students need to learn how to defend scientific propositions and communicate scientific explanations if they are to participate effectively in group discussions on scientific issues (Bybee, 2010). Furthermore, students are more likely to be able to engage in these discussions when teachers:

- model how to use questions to help students make their thoughts, reasons, and knowledge explicit and share them with the class;
- demonstrate useful ways of using language that students can appropriate for themselves; and
- provide opportunities for students to make sustained contributions to class discussions.

This type of teaching is called dialogic teaching and it is more likely to occur when teachers and students address learning tasks together, listen to others' ideas and perspectives, provide support for others to freely express their positions, and build on others' ideas to develop coherent lines of thinking and inquiry. The teacher's role during dialogic teaching is to plan and steer classroom talk with a specific educational focus.

Purpose of the Case Study
Given that many teachers face challenges in teaching inquiry-based science, they prefer to use the transmission mode to teach, believing that this enables them to cover the curriculum in the time available. Unfortunately, when this approach to teaching is adopted, students are inactive in their learning and, if they do learn,. it is often through memorisation rather than being able to link previous information to the topic presented to develop new understandings. The purpose of this case study was to investigate how one Year 5 teacher taught "What's the Matter," a unit of science from Primary Connections: Linking Science with Literacy (Australian Academy of Science, 2005). The case study focuses on how he used different visual, embodied, and language representations to capture students' curiosity in the inquiry-based tasks presented, and provided opportunities for them to investigate possible solutions and develop explanations for the phenomena under investigation. Given the key role representations play in promoting scientific literacy, the study also sought to examine the effect of the intervention on students' reasoning, problem-solving, and conceptual understandings.

METHOD

Context for the Study
The teacher who agreed to participate in this study had a science degree and had been teaching middle years' students in the school for eight years. He established the school's science club which was an out-of-class activity designed to develop and foster students' ongoing interest in science.

Inquiry-Based Science Unit
The inquiry science unit, "What's the Matter" (Australian Academy of Science, 2005), was designed to have students investigate the three states of matter (solid, liquid, and gas). It did this by providing them with different hands-on activities to observe the properties of the different states of matter, note how they change with changes in temperature, and learn how changes in the different states of matter can be represented.

The lessons in the unit use the 5Es teaching and learning model of inquiry (Bybee, 2006). This model of inquiry is designed to capture students' interest in science by providing them with opportunities to work together as they engage in different student-centred, scaffolded investigations. The teacher embedded the 5Es model into the five lessons he taught from this unit with the lessons providing students with opportunities to explore possible solutions, plan and conduct investigations, and explain and elaborate on their understandings of the phenomena they had observed. The lessons for each phase of the inquiry are outlined below.

Data Collection

The teacher was videotaped as he taught each of the five one-hour lessons in this unit. The data from these videotapes were coded according to the frequency of his use of visual, embodied, and language representations. The videotapes were also fully transcribed to specifically identify the types of teacher–student language interactions that were evident in the videos. In addition, five groups of students (three students per group) were interviewed about their conceptual understandings of the different states of matter they had been studying.

Teacher Measures

Visual Representations

Visual representations included use of a whiteboard or PowerPoint to display visual information, definitions, diagrams (e.g., graphs, tables, symbols), or to illustrate points with photographs or videos.

Embodied Representations

Embodied representations included use of the whole body and gestures to help emphasise a point or illustrate the importance of information being discussed. Information could be represented through role plays; movements of the body, hands, and head to highlight its importance; or kinaesthetic manipulatives such as shapes or the use of fingers. Embodied representations were demonstrated through visual, spatial, verbal, and motor congruency.

Language Representations

> *Basic language*: Involved making statements of facts based on knowledge of the topic.
>
> *Questions:* Included closed questions where only short, unelaborated answers were required or open questions that encouraged students to elaborate on the information that was sought. Questions were designed to stimulate recall, or clarify discussions, or promote thinking in students.
>
> *Encouragers:* Were designed to help keep students motivated and on-task and included such comments as: *Well done*; *Yes, that's right*; *I can see you're nearly finished.*

Mediation: Included paraphrasing to assist understanding, using open questions in a tentative manner to prompt students' thinking, and probing and challenging students' understanding of issues to extend their thinking. Mediation is demonstrated by such comments as: *I wonder have you thought about …?*; *On the one hand you are telling me … but on the other, you seem to be saying … I wonder how you reconcile these two ideas?*

Maintenance language: Included comments designed to ensure that the students had the resources they needed to complete tasks: *Do we have everything we need?*

Student Interviews

Semi-structured interviews were conducted with five groups of students (three students per group) on completion of the science unit. The focus of the interviews was to gauge the extent of the conceptual understandings the students had developed on the different states of matter. The interviews were transcribed and coded according to the five levels of the Structure of Observed Learning Outcomes (SOLO) taxonomy (Biggs & Collis, 1982). This taxonomy describes the increasing complexity involved in learning and includes the following levels:

1. Pre-structural level of understanding (e.g., student lacks understanding, misses the point, or uses irrelevant information)
2. Unistructural level (e.g., can recall information, names, uses terminology, makes obvious connections; performs simple instructions)
3. Multi-structural level (e.g., able to describe, classify, apply methods, execute procedure)
4. Relational (e.g., able to compare, analyse, relate, explain cause and effect, apply theory)
5. Extended abstract level (i.e., student able to transfer and generalise understandings to other topics, critique, hypothesise, and theorise; typical of intellectual maturity)

RESULTS AND DISCUSSION

The Inquiry-Based Science Lessons

The five inquiry-based science lessons that used the 5Es teaching and learning model of inquiry are presented below, along with examples of the language the teacher and students used as they engaged in different dialogic exchanges about the inquiry science unit, "What's the Matter" (Australian Academy of Science, 2005).

Lesson 1: Engage

The purpose of this lesson was twofold: Firstly, to capture students' interest and to find out what they knew about the different states of matter and, secondly, to elicit students' questions on how to identify the different states of matter. The teacher used a PowerPoint with various illustrations of the different states of matter to provide a visual representation of the topic of the lesson, to engage students in a discussion about their understanding of what matter is and how the three states – solids, liquids, and gases – can be observed. Concurrently, he introduced such terms as *particles, plasma, vapour, diffusion*, and *compressed*; terms that the students needed to be familiar with if they were to understand the different states of matter.

In order to ascertain what the students' current understandings were of matter, the teacher began the lesson by asking the students:

> *What is matter?*
> *All matter is classified into three groups. What are these three groups?*
> *What can you tell me about the particles in solids, liquids, and gas?*

While he was asking these questions, he was pointing to the PowerPoint with slides that had illustrations of the three states of matter, and asking the students what they observed as he poured salt onto a plate, poured water into a container, and held up a ball of playdough. The emphasis was on helping the students to understand how the particles in the different states of matter are compressed or diffused (using hand gestures to represent compressed and diffused). It was interesting to note that the teacher in this lesson used nearly twice as many embodied representations (63.5%) such as hand gestures and body movements as he did visual representation (36.5%) such as PowerPoint slides and pictures to communicate his ideas to the students. However, while the embodied representations were dominant, both the embodied and visual representations were congruent with the language he was using as he interacted with the students.

The following vignette provides examples of the types of questions he asked to probe students' understanding of the different states of matter and the responses the students gave:

1. T: I have poured salt into the container. *What do you notice?* (Open question)
2. S: It stays the same. We can see it. (Elaboration)

3. T: *What about the next one* (liquid)? *If I hold my hand in the water what's it going to do?* (Open question)
4. S: It's going to slip through. (Elaboration)
5. T: *If I had a solid object in the bowl what would my hand do?* (Open question)
6. S: It would not penetrate. (Reason)
7. T: *What's happening with the gas particles?* (pointing to illustration on PowerPoint slide) (Open question)
8. S: They're filling up the space (in the jar). (Elaboration)
9. T: *Salt. What about salt – solid, liquid or gas?* (Closed question)
10. S: Salt has little particles. They go all over the plate. (Elaboration)
11. T: *Liquids. What's happening with the particles?* (Open question)
12. T: *Playdough. Who thought it was a liquid?* (Closed question) (Referring to the cornflour and water mixture that has the properties of a solid and a liquid – Oobleck.)
13. S: It can be moved around like a liquid. (Elaboration)
14. S: It would move kind a like a liquid. (Elaboration)
15. T: *Need some reasons why it is a solid?* (Open question)
16. S: It can dry out and be solid. (Elaborated reason)
17. T: *We're stretching the elastic bands. What do you think is happening to the particles?* (Open question)
18. S: The particles are further apart. (Elaboration)
19. T: *Oil. Solid, liquid, or a gas?* (Closed question)
20. T: *That oil is floating on water. Why do you think it's floating on water?* (Open question)
21. S: Particles in oil are lighter than water. (Reason)
22. T: *Honey. Solid, liquid, or a gas?* (Closed question)
23. S: The particles are a little bit closer together. (Elaboration)
24. T: *What's happening?* (Open question)
25. S: The particles are a bit thicker so it goes slower (flows slowly; high viscosity). (Elaboration)
26. T: *Air. What's this one? Solid, liquid, or a gas?* (Closed question)
27. T: *Caught by air, sealed it up. What are the particles doing in air?* (Open question)
28. S: They take up the whole, entire space. (Elaboration)
29. T. *Put pressure on balloon and air moves into section and blows out. What's happening there?* (Open question)
30. S: You are compressing air in the balloon. (Reason)

It is interesting to note that of the 16 questions the teacher asked, 11 were open-ended, 5 were closed questions, and all were followed by responses from the students that elaborated on the topic. This included explanations by the students of what they had observed or reasons to explain their thinking. Normally, closed questions do not elicit elaborated responses; however, in this context where the teacher was probing the students to think about what they were observing, the students responded with more details, indicating that they realised that such responses were expected. It was also interesting to note how the students applied the correct scientific terms, *particles*, *penetrate*, and *compress*, in their responses to the teacher's questions; indicating that they were cognisant of what these terms meant.

While this first lesson was very structured with the teacher introducing key scientific concepts such as *solids*, *liquids*, *gases*, *particles*, *plasma*, *vapour*, *diffusion*, and *compressed*, which challenged the students' understandings as they observed the different states of matter, he also created opportunities for the students to work in small cooperative groups to examine such substances as stones, icing sugar, cooking oil, and honey to investigate what type of matter they represented and to explain their thinking to each other.

The following is an example of a discussion that one group had on whether the sand and the cornflour and water mix (Oobleck) was a *solid*, *liquid*, or *gas* and why they thought this.

Sand:

1. S1: Solid
2. S3: It's solid
3. S4: It's like salt (Reason)
4. S1: It's a solid because it stays in shape (Reason)
5. S4: Yeah
6. S1: It's a sold because there are many particles (Reason)
7. S3: I think sand would be a solid (Reason)
8. S1: So tiny
9. S4: All solids

Cornflour mix (Oobleck):

1. S4: It's a solid because it's like playdough (Reason)
2. S4: What do you think?
3. S1: If you squash it
4. S2: It's a liquid because
5. S1: It wouldn't be a liquid
6. S3: If you squashed it wouldn't go flat (Reason)
7. S1: It's not a gas
8. S3: Yeah but it doesn't keep its shape, because when you squash it will be like a different shape, and you can make it (Reason)
9. S4: But when you squeeze it doesn't stay (Reason)

10. S3: It won't go out your fingers
11. S2: Yeah this one's a hard one

Although this interaction represents only a few minutes of the students' discussion, it is interesting to note that nearly 50% of their total interactions on whether sand was a solid or not, involved the students providing reasons for their responses. In a similar vein, 36% of their responses on whether the Oobleck was a solid or liquid involved providing reasons for their responses; a clear indication that the students were becoming scientifically literate.

The teacher followed the small-group discussions by asking a series of questions that explored what the students understood about the three states of matter, the particles in each state, and the importance of sharing and discussing their ideas with members of their group. This was followed by a role play where the different small groups acted as particles in each of the three states of matter and explained their reasoning to the whole class. In this activity, the students were able to use each other as a resource to draw links between what they had been learning and demonstrate how they used different visual and embodied representations (hand gestures, body movements) to communicate their understanding of how the particles behaved in the different states of matter.

Lesson 2: Explore
The purpose of this lesson was to enable students to have some hands-on experiences with solids, liquids, and gases. The lesson began with the teacher asking the students to recall what matter was, how particles in solids react with each other, and the type of energy that could contribute to a solid losing its shape. When he was satisfied that the students had a sound understanding of the various states of matter from the previous lesson, they then had an opportunity to work in their small groups to investigate the properties of such objects as bread, rice, rubber bands, powders, laundry detergent, flour, honey, and air to determine if they were solids, liquids, or gases. Questions the teacher posed include:

• *What properties do they have in common with other solids?*
• *When you're looking at one grain of rice, is it hard? Is it runny?*
  We're going to get many varied answers with this.

While this was happening, the teacher moved around the different small groups to encourage the students to discuss what they had observed, make their observations explicit, and provide reasons or explanations for their positions using the appropriate scientific language. The following is part of a short interaction he had with one group:

1. T: *What do you reckon*? (Open question)
2. S: Particles are close together (Reason)

3. S: Solids are harder (Reason)
4. S: They can't be squashed (Reason)
5. T: *What is different about them, the material there?* (Open question)
6. T: *Are powders solids? Why?* You must always give me a reason (Open question)
7. S: They are like little bits of solids (Reason)
8. T: *What properties do they have in common with other solids?* (Open question)
9. S: They are rigid (Explanation)
10. T: *Any others?* (Open question)
11. S: The shape doesn't always take the shape of the container (Explanation)

This interaction between the teacher and students was only a few minutes long, yet all five open-ended questions that the teacher asked were followed by six reasons or explanations from the students about the phenomena they were observing. In so doing, the students were demonstrating that they had a clear understanding of the attributes of solids, one of the three forms of matter they were studying.

The lesson continued with the students exploring the attributes of liquids and gases and explaining their thinking to the wider class. Questions such as the following were used to help probe students' thinking, first about liquids and then about gases:

• We're going to talk about viscosity – resistance of liquid flow. *Can you give me some liquid that is high in viscosity and low in viscosity?*

The teacher used a range of hand gestures to demonstrate high and low resistance to flow, followed by pouring water into a container (low resistance) and then honey (high resistance) into another container so the students could see the difference in viscosity between the two liquids. This illustration was followed by a class discussion on examples of liquids (maple syrup, milk, chocolate sauce, orange juice) that were high in viscosity and low in viscosity.

The following is an example of the discussion that occurred in one small group when they mixed oil and water:

1. S2: For a start, with the oil I reckon …
2. S1: If we put, I think if we put the water in first and then we put the cooking oil in, then, like it would take a little while for the top of the… [Student 3 joins the group with oil and water in a cup]
3. S2: Yeah, so it would be like French dressing (Analogy)

4. T: You might have to, someone hold it steady while the other one stirs. Try and lean over. No, don't lift it out, just stir. Come over. Let it go. What's happening?

5. S2: I reckon what's happening is the particles in the oil are behaving differently because of the water, because maybe the oil may be lighter for a second (Explanation)

6. S1: Can I just try it for a second

7. S2: And because it's a new thing

8. S1: You can feel how viscous the water feels compared now with when it normally feels. [To teacher] We can feel how much viscosity (Explanation)

9. S2: Yeah, I reckon that the particles...

10. S1: And it's harder to stir (Reason)

11. S2: I reckon that the particles in the oil are behaving differently in the water because...

12. S3: The water looks different...

13. S2: They are two different liquids and they are just... behaving differently (Explanation)

14. S1: I think, this is a really stupid theory of mine, but I think maybe um that because of the viscosity in oil, the viscosity in oil, the thick, how many particles are in there, and the thickness, it's making it slower to mix into the water.... (Reason)

15. S3: But you can feel how much viscous, how much viscosity it is to stir this (Explanation)

16. S2: And it's harder to run down your fingers. Like when you rub it against your fingers, you can feel that it's greasy (Explanation)

It is interesting to note that of the 16 interactions the students participated in as they mixed the water and oil liquids to observe what happened, 8 of the students' responses were reasons or explanations as to what they had observed. In all instance, the responses were correct and indicated that the students' conceptions of the different ways these liquids interact were beginning to consolidate.

The teacher then followed this discussion with one on the final state of matter: gas.

What happens when we heat the gas? We're going to look at the circumference of the balloon when we heat the flask. The teacher had a student measure the circumference of a balloon before it was heated and then after it had been heated (the balloon was attached to a flask that was placed in hot water). The students were then asked:

- *What did you notice?*
- *What do you think was happening?*

Again, it was interesting to note that the teacher used nearly three times as many embodied representations (77%) as he did visual representations (23%) to illustrate his ideas and emphasise the points he was making. These included hand gestures to accentuate the expanding circumference of the balloon, facial and body movements to draw students' attention to the Bunsen burner (used to heat the water in the flask), and varying intonations in language to help maintain students' interest in the experiment.

The interaction that occurred above led to a discussion on the importance of a fair test when conducting experiments and the use of the acronym: *Cows Moo Softly.* In this acronym, *C* represents *Change*, the independent variable that was being measured. In this situation, it was the circumference of the balloon. *M* was the *Measure* that was used (using the centimetre tape), and *S* refers to keeping everything the *Same* (the need to keep the same balloon).

- *What are some things we have to keep the same so it's a fair test?*
- *How would we tape the tape?* (keep the tape still to ensure the test is fair)

The teacher then reminded the groups: "You're going to do a lot of explaining in your groups today (students given directions on how to work in their cooperative groups). You'll be explaining what a solid, liquid, and gas are."

The students followed this interaction by working in their small groups to complete an activity requiring them to identify some of the attributes of solids, liquids, and gases and then share their thinking during a whole-class discussion.

At the conclusion of the lesson, the teacher helped the students to reflect on what they had learned during the lesson about solids, liquids, and gases, recall the meaning of "viscosity," "properties," and "prediction," and explain the features of a fair test and why it is used in experiments. Again, the teacher actively sought to engage the students in a series of dialogic interactions that encouraged them to make explicit their thoughts, reasoning, and understandings. Opportunities were also provided for the students to engage in sustained interactions with each other about the topic under discussion because such interactions encourage students to challenge each other's thinking as they learn to defend or rebut different propositions. Mercer and Littleton (2007) observed that these types of dialogic interactions are believed to be critically important for student learning.

Lesson 3: Explain

The purpose of this lesson was to assist students to represent and explain their understandings of the three states of matter, their characteristics, and the way the different particles behave under different conditions. The teacher began the lesson by questioning the students about whether they could explain what they did in the lesson last week; what was involved in a fair test; what the anacronym *Cows Moo Softly* meant; and what they had learned from their Explore lesson from the previous week. Questions such as: *What test did you do to decide that honey is a liquid? What does a liquid need to turn into a gas?* Explain your thinking. These were typical of the types of open-ended questions he asked. He also checked for the students' understanding of the following terms that they had been introduced to previously: *independent variable, circumference, viscosity,* and *resistance.* Questions such as the following were used to probe students' thinking about the three states of matter:

> *We did a fair test last week where we changed the variables?* We measured the circumference of a balloon. *What happened to the balloon when the flask was placed in hot water? Why did the circumference of the balloon expand?*

In the other investigation with the Cows Moo Softly test. *What did we change with the viscosity test – cooking oil, orange juice, milk, maple syrup? What did we change?*

In this part of the lesson, the teacher asked a series of both closed and open-ended questions to help the students recall key information and concepts that they had learned previously in preparation for the task they would be working on in their groups.

The teacher followed this interaction with directions for the small-group activities that he had planned for the students. Included in these directions were expectations on how the students were to behave (i.e., cooperate) as they worked in their small groups, the data they needed to collect, the observations that needed to be recorded, and the requirement to report back to the class. Again, the students were reminded of what they needed to consider for a fair test as they collected their evidence to support a claim of a solid, liquid, or gas for the different samples of materials (e.g., stones, playdoh, icing sugar, cooking oil, and honey).

During the small-group activities, the teacher moved around the room monitoring what the students were doing and reminding them of the importance of finding the evidence:

- *What's the evidence, now, to support your claim of what a solid is?*
- *From the experiments last week, what evidence do you now have to say "I know what a liquid is in scientific terms," "I now know what a gas is, from the experiment"?*

An examination of the language the students used as they worked in their small groups indicated that they were able to use their scientific vocabulary correctly, such as "liquids, solids, gas, energy, viscosity, and resistance." Furthermore, they were able to offer explanations on the effects of different states of matter as shown below:

- *Viscosity is the resistance to flow. Viscous means resistant.*
- *I reckon that the particles are now ah, behave differently and they are breaking down and dissolving into a liquid substance.*
- *Um, in the solids, the particles are closer together and when you put. I'll give an example, when you put water through your hands, it slips down. If you put, for example, ice, it just stays there. And solids, when you put water into a container, it stays its shape, when you put a solid in, oh no ... to a different shape. You put a solid, it doesn't change its shape.*
- *A solid is hard and sometimes it can be soft. You can't really break a solid. The particles in a solid are all close together. Some solids can turn into liquids like ice into water.*
- *The particles in liquids are not as close together as a solid's are and can slip and slide over each other and that is why you can put your hand straight through them. If you pour liquid, for example water onto your hand, it will slide through. When you pour any kind of liquids into a glass, they take up the shape of the glass or container, depending on how much you put in.*

There is no doubt that the students had developed an understanding that some liquids can change into solids or gases when energy (heat or freeze) is applied. Again, such terms as viscous, viscosity, energy, and resistance were used, indicating that the students' scientific literacy was continuing to be consolidated.

The following interaction between the teacher and student occurred after the groups had written their description of how some liquids can change into solids or gases.

T: Can someone tell me a little bit about viscosity?
S: Viscosity is the rate of how long the liquid takes to move.
T: That's a good answer. *What else can you tell me about viscosity?*
S: It's the resistance to flow.

In the above interaction, the teacher reminded the students of the different scientific words they had encountered during their inquiry activities; words such as *"variables," "temperature," "measure," "mean," "viscosity,"* and *"resistance,"* which help to align their understandings with scientific explanations and extend their scientific literacy. He also modelled the use of these words in the scientific language he used, enabling the students to appropriate these forms of expression for themselves. This type of interaction is central to building students' scientific literacy.

Lesson 4: Elaborate

This lesson began with the teacher reminding the students that in their previous lesson, they had discussed the three different states of matter and how particles behave differently in each state. The teacher posed the following questions to help the students recap on what they had learned previously so they were ready to link this information with the tasks they were going to undertake:

- *Tell me something about particles and how different they were?*
- *What can you tell me about solids? What about liquids and gases?*
- *What are the particles doing in the different types of matter?*
- *What do you remember about a fair test?*

The students then watched a short video on the three states of matter. At the conclusion of the video, the teacher asked a series of questions to probe students' knowledge and understandings of the information presented. Questions such as the following were used to help the students recap on the information presented:

- *Tell me something about the particles and how different they were?*
- *How can salt and water be separated?*
- *Two things we're looking for today. One's called a solution and the other's a suspension. How are they different?*
- *We also looked at variables. What do I mean by variables?*

Following the video, the students moved into their small groups where they investigated what happened when different liquids were mixed (e.g., water with oil, salt, sugar). Questions they were asked to consider were whether the two liquids mixed and if not, why not? and how did the particles behave? The following vignette captures a few minutes of the discussion in one group about what happens when sugar and water are mixed:

Student Vignette

1. S2: I almost stick my finger in that
2. S1: Yeah. Do you want me to stir it up
3. S3: The sugar's staying at the bottom of the cup
4. S1: Can you hold it
5. S2: No, I've got it. Let go Olivia.

6. S3: The sugar is staying at the bottom of the cup
7. S1: But there's less sugar. You can see that there's less sugar. I want to stick my finger in there so bad.
8. S2: We can see how it behaves (puts fingers up as inverted commas)
9. S1: So what, it's not even dissolving, it's the sugar particles
10. S2: No it's dissolving
11. S1: It's, it's just hiding. The sugar particles are just hiding.
12. S2: Shall we stop?
13. S3: When the sugar particles dissolve
14. S1: It looks, it looks like
15. S3: They go back down to the bottom again
16. S1: It's dissolving but it's actually the particles are actually hiding
17. S2: Okay. Now we stop
18. S1: Where' the sugar? There's some little particles. You can see them but no one
19. S2: Oh they've all dissolved

In this discussion, the students are using many of the scientific words they had been introduced to through this unit of work on matter; particles, dissolve, dissolving, and dissolved. The teacher had previously helped the students to construct a word wall of new scientific words and these words were on the wall with the purpose of building the student's scientific literacy skills.

The teacher moved around the groups as the students continued to investigate what happens when sugar and powered milk are mixed with water. Other activities in this lesson included watching how air in balloons expanded and contracted when the containers they were attached to were heated or chilled.

Lesson 5: Evaluate
In this final lesson, students were provided with opportunities to represent what they had learned about matter, its different states, and the different ways particles in matter behave, and to bring these understandings together in a literacy product that represented their conceptual understandings. The lesson began with the teacher asking a series of questions that were designed to challenge the students to think about what they learned from their work on matter.

- *What are some of the things that we have learned from solids, liquids, and gases, or matter?*
- *What were the things that came to mind when we started this unit on solids, liquids, and gases? What did we want to find out?*
- *What causes things to melt?*
- *What did you learn about fair testing?*
- *How effective were the experiments?*

- *What did you learn about viscosity?*
- *What activity did you most enjoy?*
- *What activity did you find the most challenging?*
- *What are you still wondering about?*

This discussion was followed by the students breaking into small groups to match the different pictures and descriptions in a text to the correct solid, liquid, or gas column on a sheet of paper. The students had been given nine pictures and nine descriptions of the three different states of matter that they needed to match-up in 2 minutes. The following captures one group's discussion as they interacted together during this task.

Student Vignette

1. S1: So particles vibrate, I think that solids, particles vibrate (Basic scientific)
2. S2: Yep. What's this one?
3. S1: Ah, no. Wait. Not a shape... so I think...
4. S3: Does this look good? Move past each other, close together but move past each other (i.e., particles are close together) (Basic scientific)
5. S1: It's, yeah, liquids, um
6. S2: What's this one?
7. S1: Um solids
8. S1: Ah, gas, be compressed, squashed. Can a gas can be squashed easily, compressed?
9. S2: Yes. Gases are compressed. (Basic scientific)
10. S1: That was gas.
11. S3: This is liquid.
12. S2: Spread out ...
13. S1: Oh no, that's gas. I'm just guessing. I really ...
14. S2: Gas?
15. S2: Where's the liquid?
16. S3: Can be compressed and squashed easily. (Basic scientific)
17. S2: Gas.
18. S3: Oh, no, no, no ... I'll go get my glue.
19. S1: Liquid. Just, stick it there
20. S3: Shape, volume, leave that for last, gas.
21. S1: Ah this is liquid. Its particles are close together. (Basic scientific)
22. S3: This is gas right?
23. S1: Random direction
24. S3: Spread out to fill the space one

25. S2: Have a...
26. S3: That's a liquid
27. S2: But not shape
28. S1 That's a liquid. Have fixed shape and volume. I think
29. S3: Is this gas or liquid?
30. S1: No. fixed shape. What's fixed mean?
31. S2: To stay together. (Basic scientific)

While this activity provided little opportunity for the students to elaborate as they were matching the pictures with the descriptions on the relevant text, it is nevertheless informative to see how frequently they used the correct basic scientific language when expressing ideas or connecting information such as: "particles vibrate; gasses are compressed; particles are close together."

The students continued to work in their groups on another activity on creating large matter cards to show what they had learned about solids, liquids, and gases and to identify three properties of each. The students then played a "guessing game" where they tried to guess what state of matter is represented on each card as they listened to its properties.

The lesson concluded with the teacher asking the students to evaluate what they had learned during the science unit. The following vignette captures the teacher's final directions to the students as they were asked to reflect on their learning journey.

Teacher Vignette

1. T: Ok, you've only got one more thing to do here, and this is another evaluation... This is your self-assessment. Alright, three things I learned about matter are.... Are there three things you've learned about matter? Anyone got any ideas, or like to share what they've learned about matter? (Open question)
2. S: Matter is everything (Basic scientific)
3. T: Good. Matter, we're surrounded by matter, aren't we?
4. S: Matter is classified into three main groups: solids, liquids, and gases (Basic scientific)
5. T: Yeah, I'm glad you said the three main groups, solids, liquids, and gases, good...
6. S: That matter can decide its shape if it wants to be a solid, a liquid, or a gas (Basic scientific)
7. S: There are just three states of matter (Basic scientific)
8. T: No, there's not, there's more than three states but those are the main ones. There is a fourth state, do you remember what that one was? (Open question)

9. S: Plasma (Basic scientific)
10. T: Plasma, yeah. And I think there's actually another one, I can't remember what the other one is. The three main ones
11. S: Matter is made up of particles (Basic scientific)
12. Teacher: Good. Matter is made up of particles. So you're to do that, the three things I learned about matter. What I liked doing best. What was the best thing you liked doing about this unit on matter? (Open question)
13. S: The experiments
14. T: The experiments. Oh, we all loved getting our hands dirty and watching things go down and all that messy stuff, why?
15. S: Because it makes me feel like I'm learning things (Reason)
16. T: Yeah, and that's how sometimes you actually connect the two, don't you? By getting in there and seeing how this science connects too. We also looked at first testing. CMS, remember what CMS stands for? Cows moo softly. But what does that CMS mean? When we do that in fair testing, all the time. What's the C? (Open question)

Summary of the Teacher's Use of Visual, Embodied, and Language Representations during the 5Es Lessons
In the 5Es lessons outlined above, it is interesting to note how the teacher systematically guided the students' learning by creating opportunities and hands-on experiences that captured their interest and curiosity in the different states of matter. Concurrently, he probed and challenged their knowledge and understandings to help them explore different phenomena, explain and elaborate on what they had observed, and review current explanations as to why different states of matter behave in different ways. He achieved this by establishing a guided approach to each inquiry lesson where he introduced the problem to be investigated, used a variety of visual and embodied representations to assist understanding, asked questions that probed and challenged students' thinking, and created situations where students could work collaboratively together to perform different hands-on tasks and experiments. Finally, he assisted the students to draw conclusions based on the evidence before them. These are key processes that are central to successful inquiry learning.

Developing Deeper Conceptual Understandings
An examination of the language used by the students during their interviews indicated that most were functioning at Level 4 (Relational level) on the SOLO taxonomy where they were able to compare and relate different states of matter, analyse the evidence they collected, and explain cause and effect. Some students, though, were able to operate at Level 5, the Extended abstract level

where they could transfer and generalise understandings to other situations or topics.

The following are examples of how the students were able to articulate these enhanced understandings. For example, when asked to explain *what ice is*, one student gave the following response:

Well it's a liquid and then you put energy to it like you freeze it and then you put it turns into a solid. (Level 4: Relational)

Another student in the group augmented this comment with the following:

And then if you melt it, it goes into a liquid. (Level 4: Relational)
While a third student in the group noted: *Then if you heat that up again it goes to a gas.* (Level 4: Relational)

When asked to comment on what they enjoyed most about the work they did, the following comments were recorded:

Yep I enjoyed the whole bit about learning about matter and particles and atoms because I've never like learnt it before. This is the first year I've ever heard of it. (Level 4: Relational)
Because it involves not just numbers but even involves um accuracy and getting to see what's going to happen and try to understand and figure out and guess what's going to be the outcome of things. (Level 5: Extended abstract)

When asked to think about what makes a good representation in their science book, the following comment was received:

Yeah, you can like draw arrows to tell you like the running races we had the chair the paper, cardboard, and the part down the bottom and we drawed in an arrow like it started and then we drawed and arrow like what we were testing. (Level 4: Relational)

If you were going to teach someone about matter, solid, liquid, gas, what would you tell them, you can use your books if you want, what would tell them?

I'd tell them about what the particles look like and show them like diagrams of all these things, and solids, liquids, and gases. (Level 4: Relational)

How does keeping a science journal help you?

I actually thought it was really cool because I oh my I can forget and look back on the data and then I go so I know. Cause we've been like talking

*about we've not gone oh we've not done the runny races, next week we'll
do the runny races but that we did use that learning viscosity and I thought
looking back on the data even though I wasn't there I could think, looking
back on the data I think really, really helped.* (Level 5: Extended abstract)

The responses that the students made during their interviews clearly demonstrate the conceptual understandings and reasoning that they had developed as a consequence of their participation in the States of Matter science unit. In essence, these are ways of thinking that are critically important in helping students overcome the traditional focus on content and to learn to think and reason the way scientists do with a focus on the nature of the product achieved by such reasoning.

CHAPTER SUMMARY

The purpose of this case study was to investigate how one Year 5 teacher taught an inquiry-based science unit called "What's the Matter," a unit of science from Primary Connections: Linking Science with Literacy (Australian Academy of Science, 2005). The case study focused on how he used different visual, embodied, and language representations to capture students' curiosity in the inquiry-based tasks presented, and provided opportunities for them to investigate possible solutions and develop explanations for the phenomena they were investigating. The study also examined the effect of the intervention on students' reasoning, problem-solving, and conceptual understandings. The results showed that the students engaged constructively with their peers on the inquiry group tasks; they used the correct scientific language to make claims and discuss and compare results; and, across the course of the five inquiry science lessons, they became more adept at expressing their opinions and providing reasons and justifications for the "scientific" positions they had adopted.

ADDITIONAL READINGS

Bell, T., Urhahne, D., Schanze, S. & Plotezner, R. (2010). Collaborative inquiry learning: Models, tools, and challenges. *International Journal of Science Education*, 32(3), 349–377.
Llewellyn, D. (2014). *Inquire within: Implementing Inquiry and Argument-Based Science Standards in Grades 3–8*. Thousand Oaks, CA: Corwin.

Developing Scientific Literacy 3

INTRODUCTION

If students are to understand how science can be used as a way of thinking, finding, organising, and using information to make decisions, it is critically important that they are scientifically literate. Individuals who are scientifically literate are interested in the world in which they live, participate in discussions related to understanding scientific issues, maintain levels of scepticism about scientific claims made by others, and draw on evidence-based solutions to make informed decisions about different situations and phenomena that affect their world. This chapter provides examples of how students can be helped to relate scientific ideas to their experiences, seek answers to questions that challenge their understandings, use and interpret different representations, and engage in the discourses of science. In so doing, students learn to construct meaning about what they are learning by actively doing, reading, and writing about science in the context of socially cooperating with others.

BACKGROUND

Scientific literacy is essentially multimodal as it involves students using their scientific knowledge to identify questions about different phenomena they experience and draw evidence-based conclusions to understand and make informed decisions about the world in which they live, and the changes made to it through human activity (Rennie, 2005). People who are scientifically literate, according to Rennie, "are interested in and understand the world around them; engage in the discourses of and about science; are sceptical and questioning of claims made by others about scientific matters; are able to identify questions, investigate and draw evidence-based conclusions; and are able to make informed decisions about the environment and their own health and well-being" (pp. 10–11).

Being scientifically literate includes being able to understand and integrate multiple literacies and representational modes where language and discourse are embedded in scientific practices (Lemke, 2004). Students need to learn to make meaning of oral and written language representations if they are to be equipped to participate in public discussions about scientific issues. This involves teaching students how to read "hybrid texts" where they learn to connect the written text with mathematical expressions and calculations, diagrams, graphs, and hands-on activities such as drawing scientific representations and experiencing science in all different types of ways. In effect, students must be able to read, write, and communicate effectively if they are to engage as critical thinkers and make informed decisions about science (Pearson et al., 2010).

When you have finished this chapter, you will know:

- How students can be helped to relate scientific ideas to their prior knowledge and experiences.
- How students can seek answers to questions that challenge their understandings.
- How they use and interpret multiple representations and engage in the discourses of science.

SCIENTIFIC LITERACY

Scientific literacy, Krajcik and Sutherland (2010) argue, can be developed by: "(1) linking new ideas to prior knowledge and experiences, (2) anchoring learning in questions that are meaningful in the lives of students, (3) connecting multiple representations, (4) providing opportunities for students to use science ideas, and (5) supporting students' engagement with the discourses of science" (p. 456). The following list illustrates how these five literacy principles can be implemented:

1. *Linking new ideas to prior knowledge and experience* involves teachers connecting students' previous experiences with either real-world experiences or previous classroom experiences. Contextualising the content to be taught with students' previous experiences has been demonstrated not only to promote students' interest in what they are learning but also to help them to better understand the relevance of the science content to their daily lives (Giamellaro, 2014). For example, if

students have been learning about earthquakes and volcanic explosions, the teacher may harness this knowledge and experience by providing students with the opportunity to share and connect ideas and build on them to construct new understandings as the following illustrates:

The other day, there was a large volcanic explosion on the Italian Island of Stromboli (this volcanic explosion has been widely reported in the international press and on television). *What effect do you think it would have had on the island, surrounding sea, and people who live there?* The following visual organiser could be used to help students organise their responses: environmental, psychological, social, and economic. By working in groups of four, students could use each quadrant to record the information they had collected on each topic.

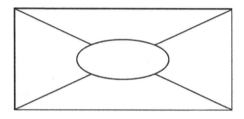

What instruments do volcanologists use to measure a volcanic eruption?
Responses may include seismometers and seismographs to measure the intensity of earthquakes, instruments that measure ground deformation and detect and measure volcanic gases and lava flows, as well as videos and cameras. Having the students collect pictures or diagrams of these instruments, report on what they measure, and share their findings with others in the class would help to ensure that all students develop a greater understanding of these instruments and their functions (Figure 3.1).

Measuring Earthquakes
You have learned about two scales to measure earthquakes – the Richter scale and the Modified Mercalli scale. What are the advantages and disadvantages of using each scale to measure earthquakes? Complete the following table and be prepared to explain your reasoning to others in your class.

	RICHTER SCALE	MODIFIED MERCALLI SCALE
Advantage		
Disadvantage		

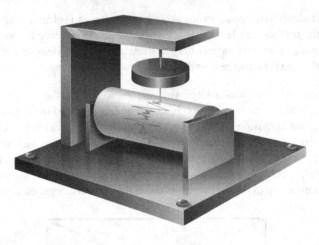

FIGURE 3.1 Seismograph.

2. *Anchoring learning in questions that are meaningful in the lives of students* is important. Students may be motivated to investigate topics if the questions posed are interesting and challenge their curiosity. For example, a question such as: *Could this devastating disaster have been prevented? What effects did it have on the population, the geography of the land, and the potential to re-build what was destroyed? What is the science behind this disaster?*

What is an earthquake?
Students could be asked to write a description of an earthquake, including information on the force and energy behind an earthquake, the movements of the tectonic plates, and the damage that occurs to the environment.

What's happening to the swimming pool in the above picture?
What's the damage? *An earthquake occurred in the small town of Shockton. The magnitude of the earthquake was 6.9 on the Richter scale. Look at the following seismogram showing the seismic waves across a short period of time. Write a description of what would be happening in the town at points A, B, and C.*

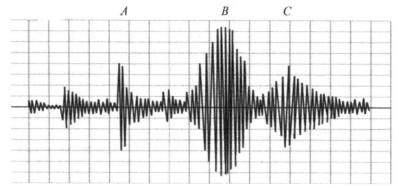

Other questions students may investigate include:

What causes soil liquefaction to occur after an earthquake? What is the effect of liquefaction on the environment?
The questions listed above could be investigated as problem-based inquiries where students work together in small cooperative groups to investigate topics that require them to collect and aggregate data and information about a topic. These types of inquiries help students to think analytically and to decide how such problems can be framed in a way that allows them to be thoughtfully examined. The advantage of working in a small group on these types of tasks

is that they have the benefit of the group's insights and analyses of the problem (Gillies, 2009).

3. *Connecting multiple representations* includes not only explicitly referring to visual elements in text and teaching students how to read and use different graphs, symbols, simulations, diagrams, and graphical illustrations, but also understanding different embodied and aural representations that support meaning making. This may include exhibitions of performance where students have opportunities to demonstrate the integration of knowledge, behaviours, skills, values, attitudes, and self-perceptions of a topic through role plays. For example, students may role-play the effect of global warming on plants, animals, water, humans, and the environment generally. The advantage of role plays is that they sharpen students' awareness of the knowledge and skills they must integrate in order to meet prescribed standards that have been previously discussed with them. Furthermore, questions from their peers in the audience help the performers to realise that they must provide reasons and justification for the information they are conveying. In so doing, they learn to be both consumers of research and producers of knowledge, further consolidating their understanding of the topic (Darling-Hammond & Snyder, 2000).

Providing students with opportunities to draw where they can sketch cells, represent particles, model the DNA molecule, or some other phenomena will help them to become active participants in learning science. It also enables them to generate their own representations to illustrate their own understandings, organise their knowledge, and integrate new and existing ideas to communicate their thinking and reasoning. Constructing their own representations is believed to allow students to articulate and refine their understanding of science ideas while enabling teachers to monitor students' conceptions as they develop, providing valuable feedback on students' conceptions or misconceptions of the topic at hand.

Narrative is another form of representation where everyday experiences and stories are shared as a way of making sense of and communicating events in the world. It is well accepted that stories are a natural way of educating students about the scientific descriptions of reality. The advantage of a narrative is that it enables teachers to link students' current experiences with the narrative, ensuring that students have a better chance of understanding the scientific explanation that is being emphasised.

Evidence is also emerging that indicates that students need to not only use and explain representations but also be able to learn new representations quickly or, in effect, demonstrate meta-representational competence (diSessa, 2004). In fact, there is a major trend towards using multiple representations including visualisations, social interaction, and written prompts to support students' conceptual development, procedural and strategic skills, and metacognition and epistemology. Furthermore, students are more likely to learn difficult concepts and principles, have better retention of what they learn, and be able to apply what they learn to problems they need to solve if the information is presented in both visual and verbal modalities rather than through one modality alone (Mayer, 2002). Opportunities to integrate information from multiple modalities enable students to engage in deeper-level learning as they learn to think with representations rather than focusing on acquiring knowledge transmitted from the teacher or some other expert.

4. *Providing opportunities for students to use science ideas* involves ensuring that students have the time and guidance to be able to apply what they have been learning in science to new contexts. For example, Danish and Phelps (2011) reported on a study where kindergarten and Grade 1 children (5–7 years) participated in a 10-week unit of work on learning about honeybees collecting nectar. Data were collected from the storyboards the children created before the unit of work and after it was completed, with the authors noting that the children's storyboard representations became more sophisticated as the children interacted with each other and sought feedback from their small groups on their representations. In short, the post-intervention storyboards very clearly demonstrated that the children had acquired a richer understanding of the content of the unit of work, and the ability of their peers to critique representations produced by others led to the production of more precise and better representations.

5. *Supporting students' engagement with the discourses of science* or teaching how to construct explanations and arguments which are key components of scientific discourse. Students do not naturally construct explanations or rebut others' ideas with clear justifications for doing so. In fact, students need to be taught how to engage in such discourse. Initially, this will involve the teacher modelling such questioning as

   (a) **Challenges**: *What are the conditions that coral need to live and grow?* (discussing environmental issues)
   When do you think this may happen given the information you have?

   (b) **Reasons are required**: *So tell me why you came up with that solution? What did you think about the work they presented? Tell us why you think that.*

   (c) **Metacognitive thinking**: *Have we covered everything? Is there anything else we still need to do? Was there anything else that you thought of in your group that you hadn't thought of before?*

(d) **Confronts discrepancies** (highlights inconsistencies in thinking): *On the one hand, I hear you saying you've covered everything, yet I wonder how you might explain leaving this out?*

(e) **Prompts**: *I'd have another look at that word. It may have another meaning that you need to consider. Have a look at that graph. I wonder if you can work out the percentage of people involved given that you know they represent a quarter of the total population?*

(f) **Helps student to focus**: *You need to consider what makes a fair test. Think of the Cows Moo Softly clue* (Change something [independent variable], Measure something [dependent variable], on the issue). *Your job is then to see if you can identify some ways in which it could be fairer.*

(g) **Tentative question**: *I wonder if you've considered if it could be done another way?* (provides another perspective to consider)

(h) **Open question**: *What are some other issues that could be considered? How would that work?*

(i) **Validates and acknowledges**: *That is a really good effort, Denver. I can see that you've worked very hard to complete the project. That's a good sentence. It's a group of words that make sense.*

Questions That Challenge Children's Understandings

Teaching students how to ask and answer questions and to provide detailed feedback that is useful to others seeking help requires a concerted effort on the part of teachers to explicitly teach and model these skills during their interactions with students. When this occurs, students, in turn, learn to model many of these questions in their interactions with each other so the responses they receive are more detailed and comprehensive than occurs with questions that do not probe and challenge students' thinking. The following are different questions that are designed to elicit different types of thinking in students that encourage them to connect the ideas they are explaining to previous knowledge and understandings. These questions have been adapted from the Ask to Think-Tel Why questioning approach that was developed by Alison King (1997).

- **Review questions**: *What does ...? Describe ... in your own words* (How does an increase in carbon dioxide affect climate change?)
- **Probing questions**: *Tell me more about ...; I don't understand. What do you mean by that?* (What do you think might happen to the ecology of the creek if the temperature of the atmosphere increased greatly?)
- **Hint questions**: *Have you thought about ...? How can ... help you?* (Have you considered the consequences that may result from allowing large passenger liners to sail into Vienna?)

- **Thought-provoking questions**: *What is the difference ... and ...? What do you think could happen to ... if ...?* (Example: What do you think could happen if there were no restrictions on selling fireworks?)
- **Metacognitive questions** (thinking about thinking): *How did you work that out? What else do we need to think of saying?* (Example: What do I still need to understand the impact of climate change on the ecology of the creek?)

In the Ask to Think-Tel Why approach, students are taught to ask each other a sequence of questions designed to scaffold each other's thinking and learning to progressively more complex levels. The advantage of this approach is that students are not only taught to focus on summarising and elaborating on information, but they are also encouraged to ask more cognitively challenging questions. These questions, in turn, help them to draw on previous knowledge and experiences and connect them to currently developing understandings. In so doing, they learn to create new knowledge and understandings about the topic under discussion.

In generating the types of questions outlined above, the questioner needs to think about how the ideas in the task relate to each other and the responder must be able to generate a response that connects the ideas together, providing explanations or rationales for his or her response. These types of responses are not unexpected because responders have been taught to provide detailed elaborations, explanations, or justifications in their response to the questions they are asked.

Question Stems and Cognitive Processes

The following are different types of question stems that are designed to challenge students' thinking by stimulating different cognitive processes. The cognitive processes are based on Bloom's Taxonomy of Educational Objectives which was designed to qualitatively illustrate different types of thinking from basic recall of information (remembering) through to creating (generating) new ideas, solutions, devising a new procedure, or inventing a new product. The purpose of this taxonomy was to highlight different ways in which individuals think from the very basic recall of information through to more complex thinking as demonstrated by creating or generating new ideas, planning new procedures, or inventing new products.

Examples of potential activities that may stimulate different types of thinking in students are provided below. These types of questions are always embedded in the context of rich and stimulating problem-solving activities that are designed to engage students' interest and challenge their thinking.

COGNITIVE ACTIVITY	SENTENCE STEMS	EXAMPLES OF ACTIVITIES
Remembering	What happened when …?	Make a timeline of ….
	Describe what happened after …	Write down the details ….
	Find the meaning of …	Make a chart showing …
	Name all the …	Make a list of all the facts
Understanding	How would you explain …?	Use a storyboard to illustrate the sequence of events
	What do you think may happen next?	Retell the story in your own words
	What was the main idea …?	Illustrate the main idea through a short role play
Applying	What are the strengths and weaknesses of …?	Construct an earthquake-proof building to illustrate your points
	In what way is … related to …?	Draw a Venn diagram to illustrate the relationship
	How would you explain …?	Make a diorama to illustrate an event
Analysing	Compare … and … to illustrate their similarities and differences	Construct a chart of similarities and differences
	How does … affect …?	How does a rise in temperature affect the sea and sea life? Graph the temperature changes and highlight the effect on sea life
	What was the turning point?	Prepare a report about the area of study
Evaluating	Which one is the best and why?	Conduct a debate that outlines the pros and cons of an issue
	Do you agree or disagree with the statement?	Write a report that supports your answer
	What are some possible solutions for the problem of …?	Outline your solutions and provide a justification for each

(Continued)

COGNITIVE ACTIVITY	SENTENCE STEMS	EXAMPLES OF ACTIVITIES
Creating	If you had access to all the resources, how could you deal with …?	Create a new product and plan a marketing campaign
	Can you create new and unusual uses for …?	Devise a way to sell an idea
	Can you develop a proposal which would …?	Outline a plan to market your proposed idea

The Discourse of Science

Engaging students in the discourse of science can be challenging as research indicates that students rarely ask each other questions unless they are taught how to do so (Howe & Abedin, 2013). One approach that can be used to harness students' interest and engagement is cooperative group work where students work together to accomplish shared goals. When students have opportunities to interact with their peers, they learn to listen to what others have to say, discuss different ideas, and give and receive help from each other. The result is that cooperative learning creates a sense of group identity where students realise that members will coordinate their actions to help and support each other, and it is this sense of group togetherness that develops that enhances student motivation to achieve. In so doing, group members work to achieve both their own goals and the goals of the group (Gillies, 2007).

The type of task that students undertake in their groups is also important because it affects how students interact with each other. When students work in small groups, they are more likely to interact with each other if the task is open and discovery based, where there is no correct answer and successful completion requires students to interact with each other and share and exchange ideas, information, skills, and problem-solving strategies. These are resources that no single individual possesses, so input from others is required. When this occurs, it is the frequency of task-related discussions that has been shown to contribute to academic gains across the curriculum (Cohen, 1994). Furthermore, the importance of arriving at a synthesis of the group members' contributions and the expectation that the group product will be presented to the wider class are also structures designed to foster group cohesion and motivate students to complete the group activity.

Encouraging Audience Participation

There is no doubt that talk is critically important to helping students shape ideas and construct new knowledge and this is more likely to happen when they have opportunities to work with their peers. Students who cooperate are more likely to participate in group discussions, provide more assistance to their peers, and demonstrate higher levels of discourse than students who do not cooperate.

Opportunities to present the outcomes of a group's task to the wider class is another way that can be used to encourage discussion among class members as they ask for further clarification on points made, challenge ideas presented, and contribute additional ideas for the group to consider. The following sentence stems capture different ways in which this may occur:

Making statements to acknowledge others' ideas

- *I liked what you said about ….*
- *I think that was a really good idea because ….*

Making statements to challenge others' ideas

- *I liked what you did but I wonder if you have thought about doing it another way to save time?*
- *I'd like to suggest that you have another look at … as there may be other issues you need to consider.*

Contributing additional ideas for the group to consider

- *Climate change is a reality so perhaps you need to emphasise that more?*

The following rubric is designed to help different groups in the audience focus on one aspect of the group's presentation and provide feedback in a constructive way. For example, one group in the audience may choose to focus on providing feedback on the overview of the study whereas others may provide feedback on the relevance of the hypotheses or the research questions. In this way, all members of the audience are cognitively engaged with the group's presentation as they know they will be called on to provide feedback.

	FEEDBACK FROM THE AUDIENCE			
CRITERIA	NEEDS TO BE CLEARER	CLEAR	VERY CLEAR	QUESTIONS RAISED
Overview of study				
Relevance of hypotheses				
Research questions				
Theory that underpins the research				
Presentation of findings (e.g., tables, graphs, illustrations)				
Discussion				
Implications				

Comments the teacher may pose for the audience to consider about the group's presentation:

- *What did you think about that?*
- *What did you think about the information they presented on that topic?*
- *Why did you decide that?*
- *Tell us what your group thought about their ideas.* (Pointing to one group in the audience)
- *Let's see if someone can help to extend their ideas.*
- *How could their presentation have been improved?*

Other ways in which students can be encouraged to engage in the discourse of science is through completing a word web. This type of activity is designed not only to provide a representation of students' conceptual understandings of a topic (e.g., genetically modified food), but it is also a tool for assessing how well the group functioned with each member of the group completing different subtasks. In this activity, the students work together to identify the economic, social, health, and environmental effects of producing genetically modified food (GM food) which they are then expected to share with the whole class. In presenting their group's word web, members of the group would be required to outline the reasons and justifications for the information they are sharing while the students in the class would be able to seek additional clarifications or challenge the information presented.

The interaction that occurs between group members and the audience often forces students to cognitively restructure the information they presented so other students can understand and use it. The process of restructuring that

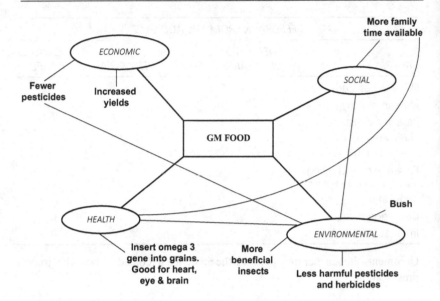

FIGURE 3.2   Word web of issues around genetically modified food (GM food).

results often requires group members to re-examine their own understandings, fill in gaps in their own learning, and provide additional clarity on the topic. In so doing, they often learn the material better than they would have if they had not been challenged to explain their points (Figure 3.2).

Linguistic Tools That Promote Student Discussion

There is no doubt that talk has a powerful effect on student thinking and the social formation of the mind with the focus of research now on how dialogic interactions or interactions with others shape the development of individual thinking (Mercer, 2008). Two of the first researchers to investigate the powerful effects of dialogic talk on students' thinking and learning were Philip Adey and Michael Shayer (2015) who focused on promoting cognitive acceleration (CA) programmes in science and mathematics in the middle years of schooling. Building on the Piagetian notion that children learn when they are confronted with information that creates cognitive dissonance, Adey and Shayer reported on a series of programmes that were implemented in mathematics and science classrooms in the middle years of schooling and found that when children are presented with cognitively challenging tasks in socially supported situations where the teacher mediates the learning, they demonstrate significant gains in cognitive development and academic achievements in comparison to age-matched peers. The cognitive gains recorded also generalised to standardised testing taken up to three years after participation in these programmes, while academic gains were recorded in science, mathematics, and English even though this last subject was not targeted in the original intervention.

Adey and Shayer (2015) believed that the success of these CA programmes could be attributed to three core elements that were embedded in each programme: cognitive conflict, social collaboration, and metacognition.

(a) Cognitive conflict occurs when children are presented with situations that conflict with their current understandings. When this happens, they must try and reconcile what they know with the new information that the current situation presents. Reconciliation only occurs when they reflect on what they currently know and how they must proceed to solve the situation they are confronting. The process of cognitive reflection enables them to re-organise and re-structure their understandings in the light of new experiences and cognitions to reduce the conflict and attain equilibrium (Piaget, 1950). Gillies et al. (2011) found that teachers promote cognitive growth in children when they use language that challenges their understandings, confronts discrepancies in their thinking, and requires them to provide reasons for their solutions.

(b) Social collaboration was the second core element that characterised successful CA programmes. In these programmes, Adey and Shayer (2015) also emphasised the importance of students collaborating with each other as they believed that intelligence is socially constructed and happens when individuals have opportunities to work with others on tasks that are intellectually challenging. When students work in well-structured collaborative groups to resolve problem issues, they understand that they must help and support each other and, in so doing, they learn to ask questions, interrogate information, and work constructively together; ways of working that have been demonstrated to promote learning.

(c) Metacognitive thinking occurs when students begin to reflect on what they are learning and what they may still need to learn. A large body of research indicates that children can be taught to engage in metacognitive processes and that they, in turn, can teach these skills to their peers. Gillies and Khan (2008, 2009) found that when teachers are taught to use specific questioning strategies that challenge and scaffold children's thinking, this leads to the development of higher-quality discourse and better reasoning and problem-solving skills in their students. Furthermore, Reznitskaya et al. (2012) noted that when teachers engage in dialogic exchanges with their students, they share the discussion, ask questions that are open to encourage students to contribute information or ideas, and provide meaningful feedback, students, in turn, learn to scrutinise both the product and processes of their discussions, and elaborate on their thinking. In so doing, they participate in the co-construction of new knowledge and understandings. In short, children can be taught to engage in higher-level thinking or metacognitive thinking and that this has a positive effect on their interactions, thinking, and learning.

Accountable Talk

Another powerful linguistic tool that can be used to promote student thinking and learning is Accountable Talk (Resnick et al. 2010). Accountable Talk teaches children how to engage in well-reasoned and logical discourse that involves being accountable to knowledge and accepted standards of reasoning within a given discipline (e.g., science, mathematics). The teacher in the Accountable Talk classroom often initiates discussion by asking thought-provoking questions that challenge students to offer detailed responses or explanations that can be accepted as valid or challenged by others in the class. During the interaction that occurs, children are taught how to interrogate other's claims, explain and justify their own positions, rebut and reconcile contradictory positions until agreement is achieved on the topic under discussion.

In Accountable Talk, the discourse is deliberative and discursive as it often involves extended and substantive exchanges between teacher and students with the teacher modelling different "talk moves" or ways of interacting to help students to clarify their thoughts, challenge and rebut different propositions, and explicate their reasoning. In so doing, students learn to engage in dialogic exchanges where they listen to what others have to say; agree or disagree in an appropriate manner; offer alternative explanations, reasons, or justifications; and augment another's argument or seek further clarification on the explanation provided.

As well as learning how to engage in dialogic exchanges with others, students also learn to be accountable to accepted standards of reasoning and knowledge in advancing their arguments. For example, students engaging in discussing phenomena in science would be expected to use the genre of argument where they make a claim, present the evidence, and outline the reasons for their proposition. Once this process is initiated, students would be expected to present the evidence for their claim while others would have opportunities to challenge or dispute this evidence with alternative explanations. The interaction that occurs is essentially dialogical with the skilled teacher guiding the exchange towards discipline-correct concepts. In this way, students learn that they must be prepared to engage in rational and logical reasoning about the presentation of evidence to support their claims. When children have opportunities to engage repetitively in this type of discourse-intensive instruction, evidence indicates that it does support learning and improved cognitive performance. Others who have investigated the powerful effect of dialogic talk on students' learning and reasoning include Mercer and colleagues (Mercer et al., 1999) who used a linguistic tool they called Exploratory Talk to help students' reason together as they worked on problem-solving tasks.

Exploratory Talk
In Exploratory Talk, students are taught how to engage in a dialogic exchange with others, where they learn how to reason and justify their assertions and opinions while challenging the ideas and opinions of others as they work together in their small groups to resolve a problem (Mercer et al., 1999). However, in order to be able to participate effectively in these types of dialogical exchanges, children are taught specific ground rules for the way they will interact.

The ground rules for Exploratory Talk during small-group discussions include:

(a) being respectful of others' ideas;
(b) sharing all information;
(c) working constructively with others;
(d) accepting responsibility for group decisions;
(e) providing reasons for decisions;
(f) challenging the ideas and opinions of others;
(g) discussing alternative propositions before making a decision; and
(h) encouraging everyone in the group to contribute their ideas and opinions.

Exploratory Talk works well when teachers play a key role in creating the classroom conditions that will enable these dialogic exchanges to occur. This includes helping students to make explicit their thoughts, reasons, and knowledge and share them with the class. It also includes modelling different ways of using language that children can appropriate and use themselves in their interactions as well as providing opportunities for children to engage in meaningful discussions with their peers.

Research on the use of Exploratory Talk indicates that it enables students to become more effective in using language as a tool for reasoning and sharing knowledge. It is also associated with higher levels of individual achievement, and significant improvements in students' capacities to solve reasoning-test problems. These results are consistent with the findings of Mercer et al. (1999) who concluded that "the use of exploratory talk helps to develop children's individual reasoning skills. It appears that even non-verbal reasoning, like that involved in solving the Raven's problems (a non-verbal test of reasoning), may be mediated by language and developed by adult guidance and social interaction amongst peers without the provision of any specific training in solving such problems" (p. 108).

Similar results were obtained by Rojas-Drummond et al. (2003) who worked with 84 children in Grades 5 and 6 in two schools in Mexico City. In the experimental school, students were trained how to use Exploratory Talk to enable them to express and share their reasoning as they worked collaboratively together on problem-solving tasks, while children in the control school were not trained to use Exploratory Talk but continued to receive instruction in their regular curriculum. The results showed that the children in the experimental classes engaged in significantly more Exploratory Talk where they utilised more detailed language in discussing the problem, reflected on different alternative explanations, offered solutions with justifications, challenged ideas, and negotiated different perspectives before deciding on a correct response. Interestingly, this enhanced problem-solving and reasoning capacity appeared to transfer to the Raven's test of Non-verbal Abilities where the children in the experimental classes performed significantly better than their peers in the control classes. The authors argued that the study confirms that language functions as a powerful tool to facilitate reasoning and problem-solving in social contexts.

Philosophy for Children (P4C)
Another dialogic tool that is used to help students' reasoning and problem-solving is Philosophy for Children. In P4C, children work together as a community of inquiry where they explore issues to try and determine what is the truth of a topic they are discussing. For example, one such issue that they may investigate is illustrated by the following stimulus question by the teacher:

T: *What is happening worldwide to our climate?*
Responses from the children may detail comments such as
S: *It's getting warmer.*
S: *There are more unusual weather events.*
S: *The sea level is rising.*
The teacher may probe the students' understanding of their comments by
     asking:
T: *How do you know it's getting warmer? What evidence do you have to*
     support your answer? What would be an example of an unusual
     weather event?

The above type of interaction continues until the participants (teacher and students) believe that they have undertaken an extensive examination of the issue and now have enough information to be able to consider the implications of any decision they make. Lipman (1988), the proponent of P4C, believed that children needed to be taught how to work constructively together to develop an inquiring outlook towards problems and issues, engage in imaginative and adventurous thinking, examine issues and ideas critically, and demonstrate a capacity for exercising independent judgement (Lipmann, 1988). Moreover, Lipmann believed that in order to be able to conduct a good inquiry, students need to learn how to pose questions to explore and analyse issues in-depth, ask questions that probe alternative propositions, explore causal connections and relationships, and pose hypothetical problems. Furthermore, questions need to also challenge students to be more self-reflective and self-monitoring.

It is through these types of dialogic exchanges that children learn to engage in reasoned argumentation with each other which helps them to clarify and sharpen their understandings and their thinking on the issue at hand (Reznitsakaya et al., 2007). Developing these types of exchanges where students are encouraged to become critical and creative thinkers takes time and involves a commitment on the part of the teacher and active engagement by students in the process.

In a corpus of studies on P4C that investigated the effects of this approach to dialoguing on students' academic, social, and cognitive abilities, Topping and colleagues (2014) found significant positive socio-emotional effects and significant gains on standardised tests in verbal and non-verbal reasoning ability that generalised to non-verbal and quantitative reasoning ability. Furthermore, these gains were maintained when students were followed up two years later even though they had not received further instruction in P4C.

Other noted benefits that occurred in the classes that received a P4C intervention included increased use of open-ended questions by teachers, increased participation of students in classroom dialogue, and improved student reasoning in justification of opinions. In an additional comment on the P4C programme, Topping and Trickey (2014) note that it does lead to enhanced reciprocal communication in the classroom between teachers and students and it clearly plays a crucial role in enhancing students' thinking and reasoning abilities that are maintained over time.

CHAPTER SUMMARY

This chapter has highlighted the importance of promoting scientific literacy among students to enable them to understand and integrate multiple literacies and representational modes, so that they may identify questions about different phenomena they experience, draw conclusions, based on evidence, and understand and make informed decisions about the world in which they live. Moreover, if students are to become scientifically literate, teachers need to be mindful of the importance of linking ideas to prior knowledge and experience, providing opportunities for students to investigate topics that are meaningful to them, and connecting multiple representations to different contexts. These literacy principles are very dependent on supporting students' engagement with the discourses of science where they learn to actively listen to each other, explain their thinking, and engage in sustained discussion about complex ideas. In so doing, students learn to interrogate others' claims, justify their own positions, and rebut and reconcile contradictory claims as they engage critically and constructively with each other's ideas.

This process of dialogic interaction is facilitated by the application of specific linguistic tools designed to promote student thinking and learning. These tools include such approaches as Cognitive Acceleration where students are cognitively challenged through the presentation of different conceptually challenging tasks that require them to collaborate with each other in order to reconcile anomalous information. Other linguistic tools that challenge students thinking and reasoning include Accountable Talk, Exploratory Talk, and Philosophy for Children (P4C). While these tools have different approaches to promoting students' dialogue, all are characterised by teachers using more high-level questions that probe students' thinking, encouraging them to analyse and speculate on ideas and challenge different propositions with evidence. Students, in turn, listen more attentively to their teacher and other students and engage in more reasoning and problem-solving interactions with each other.

ADDITIONAL READINGS

Topping, K. & Trickey, R. (2014). The role of dialogue in philosophy for children. *International Journal of Educational Research*, *63*, 69–78.
Topping, K., Trickey, S. & Cleghorn, P. (2019). *A Teacher's Guide to Philosophy for Children* . New York: Routledge.

Promoting Scientific Discourse

4

INTRODUCTION

Teachers play a critical role in inducting students into ways of thinking and reasoning by making explicit how to express ideas, seek assistance, contest opposing propositions, and reason cogently. It is well known that learning occurs when students have opportunities to interact with others where they learn to actively listen to what others have to say, reflect on their propositions, propose alternative propositions, if needed, and incorporate alternative ideas into their own understandings. This chapter introduces dialogic teaching and how teachers can use this approach to teaching to promote thinking, problem-solving, and reasoning in their students. The chapter also provides examples of how students can be taught to engage in discourse-intensive instruction that enables them to learn the genre of talk associated with reasoned discourse in science.

Emphasis in recent years has been on encouraging teachers to focus on engaging students in class discussions where they have opportunities to interact with their teachers and peers on problem-based topics that challenge their curiosity and understandings. This type of teaching is known as dialogic teaching and is designed to encourage students to be more active in the learning process by expressing their thoughts and understandings and seeking clarification on issues they do not understand. Interaction between teachers and students is critically important for student learning because it not only provides opportunities for students to demonstrate what they know, thereby consolidating their understandings, but it also enables teachers to gain a greater understanding of any misconceptions students may hold. This, in turn, allows teachers to adjust their teaching to ensure these misunderstandings are discussed and clarified.

Dialogic interactions can be promoted by the different types of questions teachers pose that probe students' understandings. For example, questions can be used to ascertain students' factual knowledge; encourage them to explain their thinking, reasons, and understandings; model ways of using language that students may choose to adopt for themselves; and provide opportunities for students to engage in sustained discussions to help communicate their understandings and clarify any misconceptions they may have. These types of questions have universal appeal because they can be used in whole-class contexts, small-group situations, or one-on-one discussions. When teachers encourage student discussion to help them explain their ideas, ask questions, explore and evaluate different propositions, and justify their own reasons, they are engaging in dialogic teaching. Dialogic teaching is aimed at achieving a common understanding through a process of structured, cumulative questioning and discussion where teachers and students build on each other's ideas that guide and prompt student thinking and lines of inquiry (Alexander, 2008).

When you have finished this chapter, you will know:

- What is involved in dialogic teaching.
- How dialogic teaching can promote thinking, problem-solving, and reasoning in students.
- How students can use different dialogic strategies to interact successfully with their peers on problem-solving activities.

DIALOGIC TEACHING

Dialogic teaching utilises the power of talk to stimulate and extend students' thinking and promote their learning and understanding. This type of talk involves teachers and students attending closely to what is discussed, asking questions to clarify issues, and building on each other's ideas to develop new knowledge and understandings. In fact, Alexander (2010) argues that this type of talk is not just any type of talk but it is underpinned by research that recognises the relationship between language, learning, thinking and understanding, and effective teaching. Moreover, it is not a single set of skills but includes a multiplicity of elements such as skills and techniques, behaviours, and dispositions that teachers draw on in response to different educational contexts, the needs of students, and the curriculum that is being taught.

Dialogic teaching is predicated on five principles designed to ensure that interaction is dialogic where teachers and students genuinely interact with each other to discover new ideas and develop new understandings. The principles of dialogic teaching require teachers to be

1. **Collective**: This involves teachers and students cooperating to tackle learning tasks together where both teachers and students listen to each other, propose ideas, and consider alternative propositions in a thoughtful, respectful, and reflective way.
2. **Reciprocal**: Teachers and students engage in dialogic exchanges where they build on each other's ideas, adding additional comments that help to further illuminate the topic until they arrive at an agreed understanding of the issue at hand.
3. **Supportive**: Students understand that they are free to express their ideas in a context that is supportive of their learning without risking disparaging comments from their peers. In this type of context, students actively work together to promote each other's learning either through helping others to understand concepts and ideas or through sharing resources.
4. **Cumulative**: Teachers and students build on each other's ideas to construct coherent and logical lines of inquiry. This involves listening to what others have to say and contributing additional information and insights that enable participants to gain a more comprehensive understanding of the topic under consideration. This type of talk is depicted by repetitions, paraphrases, affirmations, and detailed elaborations.
5. **Purposeful**: The teacher needs to ensure that the talk that emanates from the topic is well planned with the clear purpose of promoting students' understanding and learning (Alexander, 2008). There will be times when the teacher will need to guide and scaffold the students' discussions to ensure that they remain focused on the topic and not on other issues that could be distractions.

In the dialogic classroom, Alexander (2008) reports there is more discussion about talk, with teachers and students developing ground rules on how talk will be managed. Such ground rules may include:

(a) We listen to others when they are speaking.
(b) We keep to the topic.
(c) We offer constructive suggestions.
(d) We take turns speaking.
(e) We share resources.
(f) We work collaboratively as a team.

Other noticeable changes in how participants interact include:

(a) Questions tend to be designed to probe students' thinking and encourage them to analyse and speculate on ideas.

(b) Student–teacher exchanges are longer with students contributing additional ideas or information on a topic and challenging different positions that others accept as valid when the evidence has been presented.

(c) Teachers provide students with more thinking time to respond to questions.

(d) Teachers' questions are more focused and genuinely open with less emphasis on questions that cue for specific responses.

(e) Teachers spend more time prompting and facilitating students' interaction and less time on directing and controlling discussions.

(f) Teachers recognise that learning occurs in a social context where teachers and students engage in discussions where they actively listen to others' ideas, agree or disagree in a constructive manner, make suggestions, provide evidence to support a claim, and work to integrate information and ideas that helps to further illuminate the topic.

(g) Teachers and students begin to develop strategies for asking questions so they focus on one student or theme and continue to interrogate it rather than moving backwards and forwards across different students or themes.

In response to how teachers teach in dialogic classrooms, Alexander (2008) observed that students are more attentive to what others have to say and their talk tends to be purposeful with the aim of solving problem issues rather than talking at others and ignoring what they have to say. In the dialogic classroom, students use a repertoire of the types of talk that enable them to

- explain their thinking;
- analyse different points of view;
- speculate on outcomes;
- imagine solutions;
- explore possibilities;
- evaluate options;
- express their ideas;
- challenge others' propositions; and
- contribute to the ongoing discussion.

In the dialogic classroom, there is more teacher–student and student–student interaction, and interactions tend to be more detailed and extended with students providing more diverse responses with contributions of an expository, explanatory, justificatory, or speculative kind. Furthermore, more students show greater confidence in knowing how to use a range of oral skills with many speaking more readily and audibly. There is also greater participation of students who are less able in class discussions with Alexander (2008) reporting that this emphasis on talk has contributed to improvements in reading and writing outcomes for all students, including those students who are less able than their peers.

Example of Dialogic Teaching

In the following vignette, the teacher is helping the students to link what they have learned previously about the three states of matter and the particles in each state in order to prepare them for what they will be learning in this lesson. He is using a PowerPoint with slides to illustrate how the particles in the three states can be represented (i.e., dispersed or concentrated). It is interesting to see how he uses a series of open and closed questions to challenge the students' understandings of the particles and the way they are represented.

1. T: The three states of matter and the particles, how different they were. Look at those items up here. What was different about some of those items up here? We had the hot air balloon, we had a book, a cup with steam coming up. Tell me something about the particles and how different they were there. (Encouraging discussion)
2. S: Well, the solids, their particles are all rigid and...
3. T: What do you mean by rigid? What's rigid mean? (Open question)
4. S: All tight together (Explanation)
5. T: Good.
6. S: All strong (Explanation)
7. T: Keeping it nice and strong in solids. That's rigid, I can't put my finger through there. Solids are really tight together
8. S: And gas is like they're all crazy and ...
9. T: They're all crazy are they? (Closed question)
10. S: And the particles shooting round (Explanation)
11. T: Yeah, to fill up what? (Closed question)
12. S: The space of the container or the area (Explanation)
13. T: Yeah, or whatever the area, the volume inside of there is. Good boy. Do you want to elaborate a bit further on? (Open question)
14. S: Particles are running over each other (Explanation)
15. T: Yeah, running over each other and slipping and sliding so if I put my finger through, it's going to let me go through it. Good boy. We also looked at variables too, we keep coming back to variables all the time. Fair testing. What do I mean by a fair test? I need three things. And I remember this CMS thing, what's that all about? In a fair test, we have a CMS. Laura? (Open question)
16. Laura: Um, I don't, I'm not sure what the C

17. T: Who can help? CMS? Chloe M? (Closed question)
18. Chloe M: Change, measure, and same (Explanation)
19. T: Change, measure, and keep the same. So, what did we change in any of our experiments? We've done a couple now. What did we actually change in one of them? Aaron? (Open question)
20. Aaron: Um, the liquid
21. T: We changed the? (Closed question)
22. Aaron: Liquid

There is no doubt that the teacher is guiding the discussion to link students' previous knowledge and experiences with the three states of matter and the particles in each as a prelude to what they are going to learn in this lesson. Krajcik and Sutherland (2010) believe that helping students to build these types of connections in their learning is critically important to assisting students link new ideas to prior knowledge and experience if learning, in turn, is to be consolidated and conceptions developed. It is also interesting to see how he builds on students' contributions (Turns 2–4; 14 and 15) to extend their responses, enabling them to gain a deeper understanding and conception of the different states of matter.

The interactions are also very well paced as the teacher encourages the students to contribute their ideas and information to the discussion. This is achieved by his question and response sequence and his use of different "talk moves" such as paraphrasing or restating what students have said (e.g., Turns 3–7), enabling him to engage in extended discussion with the students to help them clarify and explicate their understandings. These types of exchanges are essentially dialogic-interactive where teachers and students consider different ideas and understandings together and build on each other's ideas. Resnick et al. (2010) maintain when students have opportunities to participate in these types of extended interactions, they support learning and improved cognitive performance.

While it is evident from the vignette above that the students' interactions with their teacher are more detailed, elaborative, and dynamic the more they interact, seeing how students, in turn, respond to this type of dialogic-interactive talk and use it in their small groups also needs to be explicated. In the following vignette, the teacher is minimally involved in steering the group's discussion, leaving the students to discuss the topic among themselves after an initial prompt on the different states of matter.

Dialogic Interactions in a Cooperative Group Setting
In the following vignette, the teacher sets the scene for the students to have a discussion on whether the different forms of matter are solids, liquids, or gases. In this example, the teacher is engaging in dialogic teaching where he

asks open questions, scaffolds the activity, and encourages the students in the group to discuss the topic with each other. In turn, the students engage in a range of dialogic interactions with each other:

1. T: I have my stones in my bag. Rolling stones. There they are. Ok, solid, liquid, or a gas? Can you talk to the members of your group? Have a think why they are a solid. Discuss.
2. S2: Do you think solid?
3. S1: Solid, it would definitely be a solid
4. S3: Because the particles...
5. T: Got to think why they are a solid (T. reminds students to provide reason)
6. S1: Because you can't melt them, they won't ever melt (Reason)
7. S3: Yes, you can melt them (Explanation)
8. S1: How?
9. S3: Um, magma. It's just melted rock. That's magma. All it is just melted rock (Explanation)
10. S1: And it doesn't slip out of your hand (like liquid) (Reason)
11. S2: It stays the shape (Reason)
12. S1: Because the particles are close together (Reason)
13. S3: But it can become a liquid if it wants to (Reason)
14. S2: [To teacher] Um, um what's that?
15. T: Have a think about this. Is icing sugar a solid, a liquid, or a gas? (Open question)
16. S3: It would be a solid because the particles are hard (Reason)
17. S1: [Inaudible]
18. S2: Oh yeah, um, it's it's, it's, if you put a bunch of sugar grains together, it looks like a liquid, but the actual sugar grain is solid (Reason)
19. S3: But then again it could be melted down (Reason)
20. S2: Yeah, but like in its original form, in its normal form, so solids (Reason)
21. S3: Pardon?

In the vignette above, the teacher acts more as a facilitator, encouraging the students to provide reasons for their responses (Turn 5) rather than telling them the answer. It is interesting to note that the next eight responses from the students provide reasons or explanations of whether the stones, rocks, and icing sugar are solids, liquids, or gases, providing insights into the group's developing conceptions.

Similarly, with the open question at Turn 15, the next four responses provide reasons as to why icing sugar could be a solid or a liquid. Interestingly, these responses are more detailed and elaborative than the previous ones, indicating that the students are learning to build on each other's ideas as they discuss the topic. It is also clear that the students are beginning to realise not only the importance of engaging in exchanges of ideas but also the importance of connecting ideas with evidence as occurs in critical discourse. By sharing ideas with each other, students learn how to make and defend their ideas while also receiving feedback from their peers on their thinking and the efficacy of their ideas.

STRATEGIES TO PROMOTE DIALOGIC INTERACTIONS

Teaching science involves providing students with opportunities to engage in the practice of science where they learn to acquire knowledge, skills, and dispositions that will help them to understand science and develop competencies that are important in everyday life. Students, in classrooms that provide these opportunities, learn to seek answers to issues that challenge their thinking and understandings, contest and rebut proposals from others, and actively seek solutions that are regarded as credible and viable by their peers. In a study of dialogic interactions in the cooperative classroom that involved three teachers and 17 groups of students (three or four members), Gillies (2016) found that when the teachers engaged in dialogic discussions they listened attentively to students' questions, probed and challenged their thinking, focused attention on key points, scaffolded thinking to help students connect what they had learned previously with what they currently knew, and encouraged students to explain their reasoning. In turn, the students in these teachers' classrooms were highly interactive with each other, having accumulated an inventory of linguistic tools or ways of talking that they drew upon to respond to others' questions, statements, reasons, challenges, and propositions. Gillies also noted that it was interesting to see how the students demonstrated a commitment to the group and the task they were completing by their willingness to involve others in the discussions, engage in sustained exchanges about topics, build on each other's ideas, and accommodate the ideas of others when they perceived there was value in doing so.

Huff and Bybee (2013) maintain that the most effective way to engage students in the practices of science, is to ensure the following:

1. Students have opportunities to engage in conversations with each other where they share information and ideas and reflect on what others have to say.

2. Teachers need to teach critical discourse that accentuates connections between ideas and evidence.
3. Students need to learn how to engage in argumentation where they use evidence to support their claims. Moreover, Huff and Bybee argue that as students engage in argumentation, they learn to apply their scientific knowledge in order to justify claims and identify shortcomings in others' arguments.

In order to promote the development of critical discourse and argumentation in science classrooms, the following key elements need to be built into the programme:

- **Setting expectations for classroom discussions.** This includes discussing the ground rules for classroom discussions to ensure that students understand that it is important that everyone has an opportunity to talk. For this to happen, students need to
  - listen to others when they share or challenge ideas;
  - understand that opinions and ideas are respected and considered;
  - realise that claims and reasons need to be supported with evidence;
  - recognise that alternative ideas and rebuttals need to be explicated and discussed; and
  - appreciate that the group or class need to reach agreement before deciding on an issue.

When student understand that these ground rules exist, they realise they are working in a context where they are safe to share their ideas without fear of disparaging comments being directed at them. Moreover, when students feel safe and respected by their peers, they are more likely to reach out to others by offering help and assistance and seeking help and assistance when required. In this type of supportive context, students are predisposed to embrace the challenges of mastering new information and ideas.

- **Identify learning objectives.** Students need to understand what they are expected to achieve and how they will share this information with the wider class. For example, students may be expected to construct a concept map of the relationship between animal life in a certain community; develop a food web; construct a diorama of the solar system; or enact a specific hypothesis or scenario to illustrate a scientific point or message. By being expected to share their work with their peers, students learn that they must use scientific evidence to support their claim or hypothesis if their work is to be accepted as valid. By sharing ideas in the classroom, students not only learn to make and defend claims about their understandings but are also provided with opportunities to examine their own thinking and sense making. Furthermore, as students make their ideas public, teachers are able to monitor their understandings by evaluating how they use evidence to support their claims, thereby providing opportunities for teachers to correct any misconceptions that are evident.

- **Discuss opposing evidence-based explanations or solutions.** When students are presented with information or experiences that conflict with their current knowledge and understandings, they often experience cognitive dissonance. When this happens, individuals are strongly motivated to reconcile these contradictions and search for logical coherence. In so doing, they are forced to re-examine their own points of view and reassess their validity on the basis of the new information they receive. Situations with different points of view reinforce the importance of examining alternative ideas and thinking and accepting or adjusting them in the light of this new information.

- **Observe and explain.** Providing opportunities for students to observe phenomena, explain what they see happening, and critique and justify their observations enables students to experience "doing science" rather than passively listening to a teacher telling them about what is happening. Students learn by doing where they have opportunities to explore ideas, test hypotheses, and either confirm current conceptual understandings or correct current misconceptions. Sharing and discussing these understandings with others is critically important in challenging students' misconceptions and helping them realise the need to reorganise and readjust their currently held understandings in light of feedback from others.

- **Address immature conceptions.** Teachers need to be active in ensuring that students' naïve conceptions are recognised and explanations provided as to why they are incongruent with current conceptions. Understanding why a concept may be naïve or incorrect will help students to understand why an answer is wrong and this is just as important as understanding why it is right. Huff and Bybee (2013) maintain that students' beliefs can be transformed not only by affirming understandings based on evidence but also by discussing alternative ideas and hypotheses. Students learn when they are provided with feedback that helps them to understand that their current conceptions are incorrect and that they need to consider alternative possibilities. If this discussion is dialogic, in that the teacher and student have this discussion in an open and respectful way, students are more likely to re-examine their understandings, recognise their misconceptions, and consider alternative propositions. Hattie and Yates (2014) noted that feedback is a powerful cue for learning, particularly when it helps students to understand what the criteria for achieving success is and what they need to do to achieve their learning goals.

- **Connecting prior knowledge with evidence.** It is critically important that teachers help students to connect their current understandings on scientific topics to their prior knowledge with evidence to support their claim. Ideas and hypotheses can be challenged so students need to realise that science depends on evidence that is logical and can be independently confirmed. Learning to build connections between prior knowledge and current understandings enables students to accommodate and assimilate new ideas and, in turn, construct new knowledge and understandings. The more practice students have at learning to build these connections, the more likely they are to be able to automatically engage in these processes, increasing the likelihood of learning being long term, retrievable, and transferrable to other novel and complex problems in science (Lucariello et al., 2016).

- **Practice learning.** Providing students with opportunities to rehearse and practice what they are learning is likely to lead to learning being long term and retrievable, enhancing the ability to be able to automatically apply knowledge and information to different situations when needed. Working with others provides students with opportunities to share their ideas and receive feedback from their peers on the viability of their ideas. In so doing, students learn to reorganise their thinking and communicate their ideas in ways that others will understand. When this happens, students often sharpen their own ideas, develop new understandings, and master the material better than they had previously.

- **Provide and discuss feedback.** Student learning can be enhanced when they receive regular and explanatory feedback from their teachers. Feedback is most effective when it provides students with information on their current understandings and performance in relation to their learning goals. It is also more effective when it is provided at a time when students need to receive feedback, that is, when they are actually engaged in trying to find a solution to a problem. Feedback that is delayed is not as effective because the "learning moment" has often passed and students may be less motivated to revisit a problem they had been experiencing. Feedback also needs to be discussed so students have opportunities to ask questions, clarify issues, and identify what they may need to do in order to achieve success. Feedback that is seen as fair and offers a way forward is more likely to be received than feedback that is punitive with little advice on how to overcome the difficulties the student may be experiencing.

- **Create learning situations that are challenging.** Students are more likely to engage with learning when the learning task allows them to explore topics of deep interest such as projects that they complete in conjunction with their peers. Sharing the challenges that these types of tasks can pose encourages students to take academic risks in their thinking, discuss possible scenarios to resolve issues, and share needed resources (human and material). Students also need to have enough time to think through how they will resolve these tasks if they are to stay motivated and engaged. When students are motivated, they are more likely to stay engaged with tasks because of the enjoyment and self-satisfaction they receive from doing so. Projects that students complete cooperatively have the potential to be highly motivational because they are usually designed so that input is required from all group members, that is, no one member has all the knowledge or resources to be able to complete the project without input from all members. Working cooperatively with peers encourages students to synchronise their actions with each other so the group works through tasks that are compatible with their intended aim or goal. In doing so, students begin to understand that they must pull together not only to contribute to completing the task but also to help others to do so.
- **Build interpersonal relationships.** Interpersonal relationships and communication need to be modelled and taught. Such skills do not develop by osmosis but rely on well-planned modelling of these behaviours by teachers. Students are very susceptible to respectful and courteous behaviour by teachers and will often seek to model or appropriate these ways of interacting with others including their peers. Ground rules for students' behaviours need to be negotiated prior to engaging in class activities, particularly when students are working together on projects. Project activities are designed not only to challenge students' thinking and understanding but also to ensure that all group members understand the expectations for appropriate behaviour if the members are to work constructively together.
- **Teach critical discourse.** By its very nature, inquiry science challenges students to ask questions to seek answers to problems they are confronting, so it is important that they learn how to engage in the discourses of science where they learn to seek answers to

questions in a reasoned and logical manner. Students learn to participate in reasoned and logical discussions when teachers promote discussion-based practices in their classrooms that combine thought-provoking tasks with teacher-guided discussions. In fact, Resnick et al. (2010) argue that such an approach can support the growth of both disciplinary knowledge and the capacity to engage in reasoned discussions and, in so doing, has the potential to "grow the mind." By engaging in whole-class and teacher-guided reflective discourse, students learn to explain their reasoning, make connections between ideas and evidence, and develop generalisations and conceptions.

- **Engage in scientific argumentation.** Learning to engage in the process of argumentation where students question what they are being taught in relation to what they currently know or have experienced, is certainly a central activity in helping students to realise that science is not a body of factual knowledge to be mastered. Rather, science requires students to *"think like scientists"* where they learn to engage in reasoned and logical discussions on claims, justifications, and rebuttals until a solution is accepted or rejected. Understanding how to participate in these discussions requires that students have had practice at engaging in the following sequence of activities:

    (a) **Making a claim.** A claim is usually a statement of a position. For example, *climate change is affecting weather patterns across the globe.*

    (b) **Justifying a position.** This involves the presentation of evidence such as data and an explanation of how the data support the claim. For example, *temperature records in Australia and North America show significant increases in temperature over the last 100 years.*

    (c) **Rebuttal.** This occurs when discussants challenge the evidence and suggest alternative explanations. For example, *the temperature records are incomplete (Africa) so it is not possible to assume that all countries have experienced climate change.*

    (d) **Conclusion.** This is the final part of an argument sequence and it occurs when the discussants have agreed that the evidence confirms or disconfirms the claim.

Students need to have opportunities to practice engaging in argumentation both in primary and secondary schools. In primary schools, students need to learn to distinguish evidence from opinion, listen to others' arguments and ask questions to clarify their thoughts, and learn how to construct an argument to help them to understand the phenomena they are experiencing. Examples of argument structure as it is practiced in primary schools may include:

(a) Claim: "Nicotine is a drug that affects people's health."
(b) Evidence: "Lots of people die from smoking cigarettes with nicotine."
(c) Rebuttal: "How do you know its nicotine that kills them? There may be other things involved."
(d) Explanation: "When they die and they look at their lungs, they are full of smoke and smokes got nicotine."

In lower secondary schools, students need to learn how to engage in argumentative practices by scrutinising the evidence to determine which aspects support or refute an argument; critiquing the evidence by asking questions to clarify findings; and identifying weaknesses in the evidence or claim and then proposing an alternative position. Arguments are often quite circuitous until the discussants have clarified the claim and agreed upon a solution to the problem at hand. An example of an argument may include:

(a) Claim. *The release of energy causes motion.*
(b) Evidence. *We released the pendulum and it swung back and forth.*
(c) Rebuttal. *You lifted it up and let it go so you applied force to make it move.*
(d) Explanation. *Ok. We need to apply a force to the pendulum to make it swing.*

DIALOGIC STRATEGIES FOR STUDENTS

Robin Alexander (2008) maintains that it is important that teachers who want to engage in dialogic teaching undertake to teach students how to develop a "learning talk repertoire" (p. 104). When students have a learning repertoire, they learn to propose ideas; ask questions; explore others' ideas; and argue, reason, and justify positions. Concurrently, students also need to understand the importance of developing dispositions such as a willingness to listen to others' ideas and suggestions, demonstrate cognitive flexibility in considering alternative ideas, reflect on what is discussed, and give others' time to think.

If students are to develop a learning talk repertoire, it is important that teachers organise their classes so that students have opportunities to interact, both with the teacher and each other, and that teachers continue to engage in the practice of dialogic teaching where they model and scaffold ways of interacting. When this happens, students understand how to engage in reciprocal interactions with each other where they learn to communicate and eventually master such oral skills as being able to narrate, explain, pose questions, speculate and imagine, analyse and solve problems, argue, reason, and justify, explore and evaluate, and discuss different propositions (Alexander, 2008).

Critical Thinking Skills

Students' discussions are enhanced when they know how to think critically about the topics at hand. Critical thinking involves students learning to engage in rational, thoughtful dialogue in order to clarify misconceptions, gather additional information, analyse and assess ideas and arguments, and explain results, procedures, and arguments. Importantly, critical thinking includes not only the cognitive skills associated with thinking but also the development of affective dispositions. Such dispositions include inquisitiveness about a range of issues, trust in the processes of reasoned inquiry, open-mindedness and flexibility to consider alternative propositions, honesty in examining one's own biases, and self-confidence in one's own ability to engage in reasoned discussion (Facione, 1990).

Students need to be able to think critically if they are to learn the genre of talk associated with reasoned discourse in science. One approach to achieving this outcome is to introduce students to Socratic Questioning. Socratic Questioning is premised on the belief that students can learn to think logically and rationally by engaging in the disciplined and thoughtful practice of questioning. When teachers use this approach to questioning, they promote independent thinking in their students and ownership over what they are learning. The more students practice the Socratic Questioning techniques, the more sophisticated their questioning becomes as they learn to probe issues, engage in sustained discussions on topics, and analyse and evaluate the viability of different positions.

Questions that can be used to promote students' critical thinking include:

- Questions that seek clarification on an issue. For example, *What do you mean by ...? What do you think is the key issue? Could you explain that idea a bit more?*
- Questions that challenge assumptions. *Why would this assumption have been made? What is being assumed here? You seem to be assuming ...?*
- Questions that seek reasons, justifications, and explanations. *What is an example of ...? What additional information do you think is needed? How did you arrive at this conclusion?*

- Questions that focus on consequences and implications. *What would happen if …? What could be a consequence of that …? What else could happen if that happened?*
- Questions that involve different perspectives. *How would others' respond to that suggestion? Why? What is an alternative? How are … alike and how are they different?*

Teachers who use Socratic Questioning to promote student thinking and reasoning need to be prepared to engage in reciprocal dialogues with students where they ask questions that probe students' understandings of a topic, question the evidence or the epistemological beliefs that students' advance, identify which aspects of the evidence support or refute the claim, and then help students to explain the reasons their claims can be supported or rejected. The following is an example of a teacher interacting with her students on the topic of climate change:

1. T: Climate change is in the news a fair bit lately. What do you understand about this topic?
2. S: The climate is getting hotter and we're having more bad weather with thunderstorms and cyclones. (Claim)
3. T: What evidence do you have to support your claim that it's getting hotter and we're having more variable weather? (Evidence)
4. S: The weather reports on TV are telling us about these weather conditions.
5. T: Are there other sources of information or have you been relying on the television reports? (Challenge)
6. S: I've read about it in the papers.
7. T: I wonder where the papers and the news reports get this information? (Challenge)
8. S: Scientists tell them.
9. T: I wonder how the scientists know that we are having more variable weather? (Challenge)
10. S: They study these things and they look at temperature changes over time so they can see these changes.
11. T: Let's have a look at these temperature graphs for the last 100 years. What do you notice?
12. S: The temperature is climbing. It's dipping in parts but mainly climbing.
13. T: Yes, we have some evidence that the climate is warming. What do you think may be causing this warming?
14. S: I think it may be more CO_2 in the atmosphere.
15. T: How does the CO_2 get there?
16. S: Burning coal, fires, car exhausts, gases.

CHAPTER SUMMARY

This chapter has highlighted the critical role teachers play in inducting students into ways of thinking and reasoning by making explicit how to express ideas, seek assistance, contest opposing propositions, and reason cogently. It is well known that learning occurs when students have opportunities to interact with others where they actively listen to what others have to say, reflect on their propositions, propose alternative propositions, if needed, and incorporate different ideas into their own understandings. This type of talk requires teachers and students to listen closely to what is discussed, ask questions to clarify issues, and build on ideas to develop new knowledge and understandings.

Teachers promote these types of dialogic interactions when they provide opportunities for students to engage in meaningful conversations with each other where they learn to share information and ideas and reflect on what others have to say. Dialogic interactions are also promoted when students are taught how to explain their thinking, consider different perspectives on a problem, explore alternative solutions, and evaluate possible options. This type of discourse instruction utilises the power of talk to stimulate and extend students thinking and learning. It accentuates the link between ideas and evidence, and how evidence, in turn, can be used to support claims. In short, discourse-intensive instruction enables students to learn the genre of talk associated with reasoned and deliberative discourse in science.

ADDITIONAL READINGS

Facione, P.A. (1990). *Critical Thinking: A Statement of Expert Consensus for Purposes of Educational Assessment and Instruction* (The Delphi Report). Fullerton, CA: California State University.

Topping, K., Trickey, S. & Cleghorn, P. (2019). *A Teacher's Guide to Philosophy for Children*. New York: Routledge.

Structuring Cooperative Learning to Promote Social and Academic Learning

5

INTRODUCTION

Inquiry-based science requires students to work together to investigate problems, ask questions, challenge each other's perspective, and negotiate understandings on the topic under discussion. When students cooperate, they learn to listen to what others have to say and consider their perspectives, share ideas and information, clarify misconceptions, and engage in knowledge-building practices that promote new understandings and learning. However, creating cooperative group experiences where students are able to discuss tasks in a meaningful way can be quite challenging unless students understand how they are expected to work together and what they are expected to accomplish. This chapter outlines how to structure cooperative learning so that the benefits widely attributed to this pedagogical practice are realised.

COOPERATIVE LEARNING

Cooperative learning is well established as a pedagogical practice that can be used in classrooms to promote students' socialisation and learning (Johnson & Johnson, 2002). When students work together to accomplish shared learning goals, they learn to listen to what others have to say, give and receive help, clarify differences, and construct new understandings in the context of appropriate inter-personal interactions and behaviours (Gillies, 2007). The group context enables students to interact with others and think about issues in ways that they may never have previously considered. In so doing, the information and ideas exchanged are transformed and exchanged so they become part of their new ways of knowing and doing. By engaging in reciprocal interactions with others, students learn to use language differently to explain new experiences or new realities and, in so doing, they learn to find new functions for language in thinking and feeling. The interactions that occur tend to be multidirectional as students respond to explicit and implicit requests for assistance, often scaffolding their responses to facili-tate learning in others. In fact, when this occurs in science classrooms, Ford and Forman (2015) found that the interactions that the students had with each other encouraged them to work together to collaboratively construct and critique differ-ent ideas and points of view. This participation in talk where students learn to give and take in their discussions, Ford and Forman believe is essential if productive scientific talk is to occur. Moreover, it is these dialogic interactions that, in turn, support changes to students' reasoning and scientific habits of mind.

While the knowledge-building practices of scientists are essentially social and collaborative, cooperative small-group learning emulates this by providing opportunities for students to investigate different phenomena, discuss potential hypotheses and research questions, identify the data to be collected and anal-ysed, and communicate their understandings to others in ways that are seen as logical and well-reasoned. However, many teachers experience difficulties in establishing cooperative learning experiences where students have opportuni-ties to share, critique, and evaluate possible explanations for the phenomena under investigation.

A reluctance to embrace cooperative learning may be due, in part, to the challenge it poses to teachers' control of instruction, the demands it places on classroom organisational changes, and the personal commitment teachers need to make to sustain their efforts. It may also be due to a lack of understanding of how to embed cooperative learning into classroom curricula to foster open communication and engagement between teachers and students, promote open investigation and problem-solving, and create learning environments where students feel supported and emotionally safe and secure.

Placing students in groups and expecting them to work together will not necessarily ensure that they will cooperate as some students will often defer to more-able students who may appropriate the important roles and tasks in ways that enable them to achieve their goals at the expense of other members of the group. One way to ensure that all students have opportunities to participate in groups is to structure the group so students understand that they are linked interdependently around the task and they know what they are expected to achieve and how they are expected to behave.

When you have finished this chapter, you will know:

- The benefits of cooperative learning.
- The key elements in successful cooperative groups.
- Strategies that teachers can use to promote cooperation.
- Strategies for assessing cooperative learning.

BENEFITS OF COOPERATIVE LEARNING

A plethora of studies over the last four decades have documented the benefits that students derive when they have opportunities to work cooperatively together. These benefits cover the academic, social, and emotional domains and include the development of positive and supportive interpersonal relationships, improved psychological health and well-being, enhanced communications skills, and better problem-solving and reasoning skills. In fact, it is argued that it is no longer necessary to defend cooperative learning as a teaching strategy that promotes learning as the evidence of its effectiveness is unequivocal.

The benefits of cooperative learning that have been identified include:

- Academic gains (e.g., reading, math, science, problem-solving, and reasoning).
- Positive working relationships (i.e., students learn how to develop constructive, interpersonal relationships and exhibit respectful attitudes to others).
- Less need to discipline (i.e., the group acts as a model of how to behave for members to adopt).
- Higher levels of self-esteem (i.e., feedback from peers can be positively affirming).
- Promotes acceptance of others (i.e., contact with others in a supportive environment promotes a better understanding of an individual's needs).

- More inclusive language (i.e., the use of "we" and "us" to denote a sense of group inclusion).
- More detailed explanations (i.e., more detailed explanations are associated with learning gains).

Advantages of Small, Cooperative Group Instruction

Lou and colleagues (1996) have identified numerous advantages to small, cooperative group instruction. These include:

- Emphasis on the diversity of instruction rather than uniformity where all students are taught the same way. Teachers can adjust their teaching to accommodate the more-able and less-able students without drawing attention to the specific needs of individual students.
- More time for peer learning and teacher assistance. Peers are often quite sensitive to what other students do not know and will explain any misunderstandings in a way that others will understand.
- Greater flexibility for teachers to adjust learning objectives. Even in mixed-ability groups, teachers will often set different learning objectives for different students to ensure that the more-able students are challenged while the less-able students are assisted to complete their tasks or goals.
- Repetition for less-able achievers. The opportunity for less-able achievers to participate in small groups where they have opportunities to hear and see what their peers think about a topic, often helps to promote their understandings and learning.
- Students have opportunities in small cooperative groups to orally rehearse material, explain it to others, discover solutions, debate its merits, and discuss procedural issues.
- Opportunities to promote higher-order thinking skills. As students spend more time talking and thinking about topics, listening to the ideas of others, and reflecting on them, they often develop better reasoning and problem-solving skills.
- Motivation to learn information. When students feel included in a group and that their efforts are appreciated by their peers, they are more likely to remain engaged in the activity and motivated to learn.
- Opportunities to develop social and communication skills. Small groups provide opportunities for members to learn appropriate ways to behave and interact with each other.

Types of Cooperative Learning Groups

Johnson and Johnson (2009) identified three types of cooperative learning: formal, informal, and base groups. Formal cooperative learning involves students working together, often for one class period over several weeks to achieve shared learning goals and complete specific tasks and assignments. Formal cooperative learning involves the class teacher in pre-planning what the groups will be doing and how they will work together. This includes the types of tasks the groups will be asked to undertake and how the subtasks within the group will be allocated so all group members are meaningfully involved in contributing to the group's goal of completing their assigned task.

Formal cooperative learning also includes discussing the importance of appropriate social behaviours so that all members feel they can express their opinions within the group without being sanctioned or criticised by others. It is important to discuss the expectations of behaviour that the teacher and students hold for each other as part of the process of establishing the group, so group members know how they are expected to behave and how, in turn, they expect others to behave. The teacher's role in establishing this type of group is to monitor the members' progress towards achieving their group goal and to intervene to improve task work and teamwork when it is appropriate to do so. The teacher is also responsible for assessing student learning (formally or informally) and helping the group to process their effectiveness. In short, what they have managed to achieve, what they still need to do to achieve their goal, and how they will do this.

Informal cooperative learning, on the other hand, involves students working in small groups for a few minutes to help members process what has been taught, to think about a particular question, to assist the teacher to identify and address any misunderstandings about the content or processes associated with how the group is working together. Informal cooperative learning may consist of pairs of students or larger groups (three to four members) who collaborate around a task, often for only a few minutes throughout the lesson. The teacher's role in establishing informal cooperative learning groups is to discuss the purpose of this type of group at the start of the lesson, monitor the group's progress, and allocate time on completion of the lesson to discuss what the groups have achieved during their informal cooperative learning activities.

Base groups are long-term, heterogeneous, cooperative learning groups with stable membership. The members' primary responsibilities are to

(a) provide one another with support, encouragement, and assistance in completing assignments;
(b) hold one another accountable for striving to learn; and
(c) ensure that all members are making good academic progress.

Typically, cooperative base groups are heterogeneous in membership, especially in terms of achievement, motivation, and task orientation. They meet regularly for the duration of the class (Johnson & Johnson, 2009).

KEY ELEMENTS IN COOPERATIVE LEARNING

Research has identified five key elements that need to be present in groups in order for members to cooperate (Johnson & Johnson, 2002). These elements are

(a) Positive interdependence; the perception that group members are linked together in such a way so that all must succeed if the group is to achieve its goal.
(b) Promotive interaction involves actively assisting others' learning.
(c) Interpersonal and small-group skills.
(d) Individual accountability.
(e) Group processing.

The Five Key Elements in Cooperative Learning

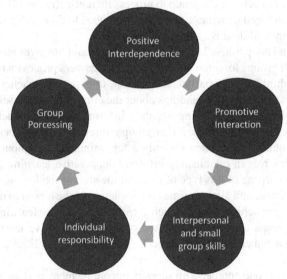

(a) **Positive interdependence** exists in small groups when students understand that they are not only required to complete their part of the task but that they are also required to help others do likewise. The technical term for this dual responsibility is "positive interdependence" and it is the most important element in cooperative learning. Positive interdependence exists when students perceive that they cannot succeed unless others also succeed and they must synchronise and coordinate their efforts to ensure that this occurs. Group cohesiveness develops as a direct result of the perception of goal interdependence and the perceived interdependence among group members. In essence, they learn that members of the group either "sink or swim together" and that each member's efforts are essential if the group is to succeed in completing the task. When positive interdependence is not established in groups, members may collaborate in an "ad hoc" way or they may choose not to cooperate; working instead on their own specific goals or tasks.

Positive interdependence may by structured by asking group members to

1. Work towards achieving a group goal or task (group goal/task interdependence).
2. Ensure that each member can explain the group's overall answer (learning goal interdependence).
3. Fulfil assigned role responsibilities (e.g., resource manager, timer, scribe).

Other ways of structuring positive interdependence may include having common rewards (reward interdependence), shared resources (resource interdependence), or a division of labour so different students accept the importance of investigating specific sub-topics individually and then sharing their expertise with others in the group (sometimes called expert groups).

Johnson and Johnson (2009) found that when positive interdependence is established in a group, two important psychological processes occur. The first involves members allowing one member's actions to substitute for the actions of another. This occurs when one member undertakes an action that other members of the group accept as an action they see as important to the group. The second psychological process is being open to the influence of others and willing to accept their ideas as valuable. When these two processes are evident, the members become psychologically interdependent with members realising they need to work together, be open to assisting others, and contribute their ideas to ensure the group completes its task or achieves its goal/s.

(b) **Promotive interaction** involves students encouraging and facilitating each other's efforts as they work together. This may include students orally acknowledging each other's efforts, providing information and resources to assist understanding, and offering constructive feedback to improve performance on a task. Promotive interaction is facilitated when students work in small groups where they can hear and see what others are saying and doing, enabling them to develop a better awareness of the difficulties others may have and provide assistance as needed. Students are often very perceptive of what others do and do not know and will often provide help not only when it is requested but also when it is not explicitly requested. In this sense, the group experience assists students to be more "in tune" with others' needs. When group members demonstrate these types of behaviours, students are more likely to feel accepted and valued, less anxious and stressed, and willing to reciprocate and help others in turn. Furthermore, as students work together, they are more likely to get to know each other personally and this forms the basis for developing caring and supportive relationships.

Research also indicates that students work better together when there are no more than four members in each group and there is a balance of male and female students. Groups larger than four often lead to some members being overlooked by others as they struggle to join the discussions because they have difficulty hearing the discussion or viewing the group's task sheet. Group composition is another issue that needs to be addressed because in gender imbalanced groups (e.g., more males than females), the male members will often dominate the discussion to the detriment of the female member. Interestingly, in groups where there are more female members than males, the female members have been shown to work hard to include the male member, often to the detriment of their interactions with each other. In short, it appears that both males and females are more likely to interact with each other and promote each other's learning in gender-balanced groups.

Specific strategies to promote interaction among students include:

(a) providing others with efficient and effective help and assistance, exchanging needed resources;
(b) providing others with feedback in order to improve their continuing performance on tasks;
(c) challenging others' conclusions;
(d) advocating efforts to achieve mutual goals;
(e) influencing others' efforts to achieve mutual goals;
(f) having faith and trust in others;
(g) being motivated to strive for mutual benefits; and
(h) feeling less anxiety and stress.

(c) **Interpersonal and small-group skills.** It is critically important for the success of cooperative learning that the social issues regarding the interpersonal skills and the small-group skills that students will be expected to demonstrate are clarified before students begin to work together. Gillies and Ashman (1998) found that when students were trained in how to use these skills, they demonstrated more cooperative behaviour, provided more help to each other, and used more inclusive language than peers who had not been taught these skills. Some of the benefits that were evident in the trained groups included more autonomy and more successful learning outcomes. This may be, in part, because as students learned to interact appropriately with their group members, they felt more supported by their group members and were more willing to reciprocate in kind. There is no doubt that social support tends to increase group cohesion and sense of purpose which, in turn, affects the pressure to be a productive group member.

The interpersonal skills that facilitate communication among group members include:

(a) Actively listening to others when they speak.
(b) Stating ideas freely without fear of being sanctioned.
(c) Considering the perspectives of others when they express an idea or opinion.
(d) Accepting responsibility for one's own actions or behaviours.
(e) Providing constructive criticism about an idea, topic, or decision.

The small-group skills that facilitate group communication include:

(a) Taking turns with expressing opinions or sharing resources and roles.
(b) Sharing tasks so one individual is not overloaded with work.
(c) Making decision democratically so no one dominates, and all have a say.
(d) Clarifying differences or misconceptions.
(e) Engaging in democratic decision-making.

In order to be able to cooperate, students must

(a) get to know each other and trust each other;
(b) communicate accurately;
(c) accept and support each other; and
(d) resolve conflicts constructively.

Examples of Interpersonal and Small-Group Skills

SKILL	LOOKS LIKE	SOUNDS LIKE
1. Listening	Eye contact	Yes; I see; Ah! Mm!
2. Stating ideas clearly	Scan group, face group	I think…
3. Constructive criticism	Eye contact	I liked … but have you thought of …
4. Accepting responsibility	Scan group, face group	I-statements
5. Sharing tasks	Pass the materials, jobs	Have we all got something?
6. Taking turns	Eye contact, facing group	I've had my go…
7. Understanding others	Appropriate facial gestures	Do you mean …? Are you saying …?
8. Clarifying differences	Eye contact, face group	I'm not sure I understand …?

Skills that facilitate interpersonal communication

1. Body language is very important because non-verbal body language can be very effective in communicating with others. For example, eye contact tends to denote interest, an open posture indicates that the individual is open to listening to others, and gestures and facial grimaces can be used to highlight the importance of a message.
2. Verbal encouragers are used as ways of communicating that one is interested in what is being said. Verbal encouragers may include: *Mm!*, *Ah! Sure*, and *Yes*. These encouragers are designed to encourage the speaker to continue talking.
3. Open questions are used to encourage individuals to talk more openly about an issue or problem. Open questions usually begin with: *How, What, When, Where*, and *Why*. Other ways of encouraging an open discussion include: *Can you tell me a bit more about what happened?*
4. Paraphrasing involves listening carefully to what has been said and then restating the main idea and repeating it to the speaker. For example, if someone makes a comment about having had a long and tiring day, the paraphrase may be: *It's been a tiring day for you. Is that correct?* A check on whether the paraphrase is correct is always

undertaken to ensure the paraphrase is correct. A paraphrase communicates to the speaker that the listener is *in tune* with what has been said.

5. Summarising main points of the story or interaction. This is often undertaken to ensure that the listener can recall the main points or gist of the story or interaction. A summary may include: "So you said (a)..., (b)..., and (c).... Is that correct?"
6. Empathic identity skills:
    You must have been scared....
    Sounds like you've had a rough trot....
7. Clarifying misperceptions: *I'm not sure I understand what you've said. Can you explain it to me again, please?*
8. Expressing a point of view assertively: *I think you've got a good point there, but I think that it might be better if we....*
9. Tentatively offering suggestions:
    Have you thought of...?
    Maybe you could do it this way?
10. Self-disclosure: *We all make mistakes. I know what it was like for me when....*

(d) Individual accountability or personal responsibility includes
    (a) 7014210-6311903. Individual accountability003. Individual accountabilitybeing responsible for completing their part of the task and facilitating others to do likewise;
    (b) reporting to the group on their progress;
    (c) reporting on the group's progress to the whole class; and
    (d) being rewarded (e.g., receiving bonus points) or acknowledged on the basis of all group members completing their tasks/goals.

When individual accountability or personal responsibility is evident, group members realise that others are also contributing to the group's goals. This, in turn, helps to create a sense of group cohesion and motivation to work together as members realise the importance of their contributions to the group's goals. Johnson and Johnson (2009) found that individual accountability or personal accountability increases the effectiveness of a group and the work members do by ensuring that everyone contributes to the group's goals.

Strategies to promote individual accountability include:

- Group members develop specific criteria for assessing each member's contributions.
- Group members may be assigned individual tasks to complete which they share with the larger group when these tasks are completed.

- Group members are assigned different roles which they need to undertake if the group is to achieve its goal.
  (e) **Group processing and reflecting.** These are processes that are critically important for student learning as they allow members to discuss how well they are achieving their goals and maintaining effective working relationships. Johnson and his colleagues (2009) suggested that both teacher-led and student-led discussions promoted greater success in problem-solving and achievement gains than students involved in cooperative learning who did not follow up with processing their experiences in groups.

Group processing involves reflecting on a group's session through

  (a) Describing members' actions that were helpful and unhelpful.
  (b) Making decisions about what actions to continue or change.

Involving students in reflective processes promotes metacognitive abilities which impact on students' abilities to provide supportive feedback and to more frequently engage in appropriate social behaviours. One important and potentially reinforcing aspect of group processing is that it can provide an opportunity for group members to celebrate their success.

Strategies to Promote Group Processing
The following strategies may be employed to promote successful group processing:

- Considering the format required and the information that will be obtained, approximately 10 minutes at the end of a session is devoted to students' identifying what they have achieved and what they still have to learn. A number of formats are available to assist this process. Commonly, questions are posed to encourage a "yes/no" response. For example, *Did we encourage one another to contribute to the discussion?* Where deeper processing might be required, students are encouraged to consider the quality of behaviour across a continuum. For example, in response to the statement, *We encouraged one another to contribute to the discussion*, students choose along a continuum from *Strongly agree* to *Strongly disagree*. Some formats also allow for student comments.
- Processing is linked with a problem-solving metacognitive strategy. For example, groups are expected to follow a procedure for executing a task. For example, *What is the problem? What are we going to do? How are we going? How did we go? What might we have done better?*

- Creating quick representations (e.g., a graph) to show group effort, what has been achieved, and what still needs to be achieved.

At the end of a unit of work or project, students as individuals can be encouraged to consider how "I" contributed to the productivity and welfare of the group.

STRATEGIES FOR CONSTRUCTING COOPERATION IN GROUPS

The teacher plays a key role in establishing cooperative learning in the classroom to ensure its success. This includes being aware of the importance of how to structure these groups, including the size and composition of the groups, the activities they will be expected to undertake, the responsibilities of both members and groups, and the teacher's role in monitoring the group's processes and outcomes.

1. Groups of three or four persons are better than larger groups, possibly because if groups are too large, they tend to be less personal and students will not participate. In small groups, everyone has an opportunity to be involved, whereas in larger groups, it is harder for members to hear and see what is being discussed. In larger groups, one or two members tend to take over responsibility for completing the task, leaving others on the sidelines with minimal involvement.
2. Mixed-ability groups are better than same-ability groups. Lou et al. (1996) found that low-ability students learn significantly more in mixed-ability groups, possibly because they benefit from the tutoring or assistance they receive from their more-able peers who tend to provide help not only when it is requested but also when they detect it is needed. In contrast, mixed-ability students learn significantly more when they work with students who have similar abilities to their own, possibly because they are more verbally active, giving help and receiving help from each other. Interestingly, high-ability students learn equally well in mixed-ability or high-ability groups.
3. The gender composition of groups is important. Groups tend to work well with all members participating when they are gender balanced; for example, two males and two females. Groups also work well when they are all male or all female groups as their discussions are often quite animated when they are pursuing common interests. Difficulties appear to arise when there is an imbalance in genders; for example, more males than females. In this type of group, the female member tends to be

overlooked while the males converse among themselves. Interestingly, females are also disadvantaged when there are more females than males; for example, three females and one male. In this type of group, the females spend a disproportionate amount of time trying to engage the male in the discussion rather than interacting among themselves.

4. Friendship groups may be better with adolescents. Teachers need to be mindful of the friendships' adolescents have and the common interests they may share. On certain tasks, it may be more appropriate to have all male or all female groups or groups where students share common interests, such as music or sporting groups.

5. Case-based learning (CBL) involves students collaborating to find solutions to real-world problems. The advantage of CBL is that it is a student-centred approach to learning that requires students to use their reasoning and problem-solving skills in order to find a solution to the problem under discussion. According to the National Centre for Case Study Teaching in Science, cases should be authentic and based on real-world scenarios, tell a story that captures students' interest, be aligned with learning goals, and be able to challenge students' thinking and reasoning. In classroom settings, it is critically important that students have opportunities to discuss their investigations with others, where they share their developing understandings, critique and rebut suggestions, and develop solutions in the context of a supportive and encouraging environment (Herreid, 1994).

STRATEGIES FOR ASSESSING COOPERATIVE LEARNING

Assessment plays a key role in educational accountability. Being able to assess students' progress in cooperative learning is critically important, particularly because responsibility for learning is transferred to the group with the teacher acting as a "guide on the side" rather than an instructor in the learning process. When students work in small, cooperating groups, teachers need to assess not only how students are managing the learning process but also what they are achieving (learning outcomes). While students derive clear benefits from being able to work cooperatively with others, effective assessment practices require that these benefits are documented so teachers can share them with parents, students, and reporting authorities. This process also enables teachers to reflect on their own teaching practices to determine what they may need to change or adjust to promote improvement in their students' learning.

Teachers often use a variety of approaches to collect information on how their students are managing the learning process, from observing students' learning, collecting work samples, listening to students' discussions as they work together on a topic, having students report to the wider class on a topic they have been researching, and encouraging students to keep a journal of their learning experiences. Other types of formative assessments may include peer assessments where peers assess the process of learning such as how they worked together and the quality of the work they achieved. These types of informal approaches to collecting information on how students are managing their learning are generally referred to as formative assessments. In this sense, formative assessments are designed to provide information on the ongoing teaching and learning process to determine its effectiveness.

A second type of assessment is summative assessment which is designed to measure what students have learned at the end of a period of work or instruction. In cooperative learning, summative assessments provide information on what the group has accomplished, and they may include the presentation of a group product (e.g., report, diorama, model), a group performance (e.g., role play, debate), and criterion-referenced assessments to determine if students have achieved pre-determined criteria for completing the task. These types of assessments need to have the following features:

- Involve real-life situations
- Are intellectually challenging
- Assess students' higher-order thinking
- Are motivational and educational
- Are criterion referenced
- Encourage student self-reflections
- Are collaborative

Another form of summative assessment is case-based learning where students collaborate to solve a problem that they are likely to encounter in real-life learning environments. Case-based learning usually involves problems that are complex and ill-defined with no set procedures to follow. In order to solve these problems, students need to have access to a range of resources that will enable them to explore the problem in-depth and have access to the critical insights of others as they discuss possible solutions to the problem at hand. Multiple criteria are usually used to assess case-based learning. These criteria may include:

- Identification of suggested solutions.
- Consequences that may apply.

- Identification of the best solution and reasons why, including utilising well-developed arguments for the choices made.
- Metacognitive thinking (i.e., combines different thinking processes).

The research on assessing cooperative learning indicates:

- Both formative and summative assessments can be used to assess student learning.
- Students learn more when formative assessments are criterion based.
- Criterion-based assessments are designed to help students understand what they need to do to achieve specific outcomes.
- Teachers need to provide feedback to students that is specific and helps them to understand what they have accomplished and what they still need to do.

CHAPTER SUMMARY

This chapter underscores the importance of structuring cooperative learning in groups so members understand what they are expected to accomplish and how they are expected to behave. The key elements that need to be present in groups in order for members to cooperate include: positive interdependence (members actively work together to complete the group's goals); promotive interaction (members encourage each other's efforts); interpersonal and small-group skills (members demonstrate appropriate social behaviours); individual accountability (each members needs to be responsible for completing their part of the task); and group processing (members reflect on what they have achieved as a group).

Other issues that need to be considered when establishing cooperative learning in groups include the size of the group (no more than four members), the gender composition (gender-balanced groups are preferable), the ability composition (members appear to learn more in mixed-ability groups), and friendship groups may be better for adolescents. Case-based learning where students use their reasoning and problem-solving skills to solve real-world problems are highly motivational because of the challenge they pose.

Assessment practices are also critically important as a way of monitoring both the process of learning and the outcomes achieved. Assessments may include both formative and summative assessments. Formative assessments are designed to provide information on how students are managing the learning process so support can be provided if needed while summative assessments are designed to determine what has been achieved.

ADDITIONAL READING

Gillies, R.M. (2007). *Cooperative Learning: Integrating Theory and Practice.* Thousand Oaks, CA: Sage.

The Structure of Observed Learning Outcomes (SOLO) Taxonomy

6

Assessing Students' Reasoning, Problem-Solving, and Learning

INTRODUCTION

This chapter discusses how the Structure of Observed Learning Outcomes (SOLO) taxonomy can be used to assess student reasoning, problem-solving, and learning. The SOLO taxonomy describes the increasing complexity involved in learning and includes levels that range from (1) pre-structural level of understanding (e.g., student has no knowledge or disconnected bits of information); (2) unistructural level (i.e., student can recall, use terminology, name, and perform simple instructions); (3) multi-structural level (i.e., student can describe, classify, and execute procedures); (4) relational level (i.e., student can compare, analyse, relate, and explain use and effect); and (5) extended abstract level (i.e., student able to transfer and generalise understandings to other topics, critique, hypothesise, and theorise – typical of intellectual maturity). The focus in this chapter is on how the SOLO taxonomy has been used as a way of gauging the complexity involved in students' reasoning, problem-solving, and learning during inquiry science activities.

When you have finished this chapter, you will know:

- The five levels of cognitive processing that are part of the SOLO taxonomy.
- Key words and processes that are the intended learning outcome at each level.
- Examples of how the SOLO taxonomy can be used to identify students' learning competencies.

THE SOLO TAXONOMY

The Structure of Observed Learning Outcomes was first proposed by Biggs and Collis (1982) in their book, *Evaluating the Quality of Learning: The SOLO Taxonomy*, and is based on the outcomes of academic teaching. This taxonomy describes the increasing complexity involved in learning and includes levels that range from (1) the pre-structural level of understanding to (5) the extended abstract level (i.e., student is able to transfer and generalise understandings to other topics, critique, hypothesise, and theorise, typical of intellectual maturity). Table 6.1 provides information on the five levels and examples of the responses the students generated.

The SOLO taxonomy is not a developmental theory of learning such as Piaget's (1950) Theory of Cognitive Development, which proposes that children proceed through different stages in an irreversible sequence from pre-operational to concrete to formal. Piaget argued that the sequence of development was hierarchical and that the stages followed one after the other from the sensorimotor stage (birth to about 18 months), pre-operational (4–6 years), early concrete (7–9 years), middle concrete (10–12 years), and concrete generalisations (13–15 years) until the formal operations stage (16 years plus). Furthermore, Piaget believed the stages and sub-stages were stable so once individuals arrive at a particular developmental stage, they think in alignment with that stage. For example, a child who has the ability to think in a way that is characteristic of an early concrete level (7–9 years) would be expected to demonstrate this type of thinking across all problem-solving tasks and school subjects.

TABLE 6.1 The SOLO Taxonomy and Examples of the Five Levels of Increasing Complexity in Students' Language

LEVELS OF COMPLEXITY IN THE SOLO TAXONOMY	EXAMPLES OF QUESTIONS ASKED AND STUDENTS' RESPONSES
Level 1: Pre-structural level (e.g., student lacks understanding, misses the point, or uses irrelevant information)	Single word response, no elaboration or connection to topic being discussed. Student may also comment with words such as "don't know," "can't do it," or "it's dumb."
Level 2: Unistructural level (e.g., can recall information, name, use terminology, make obvious connections; perform simple instructions)	Interviewer: What do you like learning about science? What do you enjoy? Student: You get to see how you're doing it. You actually get to experience experiments. (Level 2: Unistructural) Interviewer: Which experiments did you like? Student: Oh! the runny races. (Level 2: Unistructural)
Level 3: Multi-structural level (e.g., able to describe, classify, apply methods, execute procedure)	Interviewer: What's the point of having a race between orange juice and milk and water and maple syrup? That's a silly thing to do having a race. What would you tell them? Student: That when you have the race that you're measuring the most viscous and less viscous liquid. (Level 3: Multi-structural)
Level 4: Relational (e.g., able to compare, analyse, relate, explain cause and effect, apply theory)	Interviewer: Hummm! So how would you tell me about the way the particles work in those different things? Student: Well, in the solids they're all like packed tightly together so you still keep their shape… so they can't move very much like over themselves, like a liquid. (Level 4: Relational)
Level 5: Extended abstract (e.g., able to perceive structure from different perspectives, transfer and generalise understandings to other topics, critique, hypothesise, and theorise)	Interviewer: You can say anything, ok, all right I'll just see if there's anything oh, I'm just wondering about your science journal ok, did you enjoy doing your science journal? Student: Well you could record data and if you like if you were struggling you could like see if you wanted to get some feedback on the answers. (Level 5: Extended abstract) Interviewer: Right so what do you think, you know how some people say what is the catalyst, now what is the thing that changed for you, what do you think changed? Student: I liked figuring out um how the particles, like kind of worked together cause I didn't know that solid particles vibrated and the liquid particles slid over the top and I didn't know the gas particles went everywhere so I thought that was really good. (Level 5: Extended abstract)

However, from their review of hundreds of students' responses in elementary, high school, and college across different subject areas, Biggs and Collis (1982) found that the assumption of stage theory was not supported. In fact, they found that it was possible for students to operate at one level in one subject and at another level in a different subject. They further argue that when the focus is moved from a student's developmental level to examining a response to a particular task (observed learning), it is easier to understand how a student may think and function at a concrete operational level on one task and at a pre-operational level on another task, depending on the student's understanding of the task, motivation to engage with it, learning strategies that are implemented, and the effectiveness of the instruction received.

In contrast to the Piagetian Theory of Cognitive Development with its focus on stages, Biggs and Collis (1982) argued that the SOLO taxonomy focuses on what an individual has achieved from instruction. Student achievements are very dependent on a combination of effective teaching and student intentions. Teachers are more likely to be effective when they actively seek to engage students' interests in a topic; plan learning tasks so that students' understand what they are expected to do and achieve; utilise learning strategies that cater for the needs of all students (i.e., set tasks that are appropriate for the child's style of learning and draw on prior knowledge); and provide feedback that helps students understand what they have achieved and what they may still need to do. When students feel supported in this way, they are more likely to engage with learning that challenges their understandings, knowing that they are not likely to be rebuked if they do not understand the task. Learning processes are complex and very dependent on a student's prior knowledge, developmental stage, and working memory; the strategies the teacher uses to teach different concepts (i.e., scaffolding, explicit teaching); the learning strategies the student has acquired; and the student's motivation to learn.

The SOLO taxonomy is based on the learning outcomes that students achieve as a result of a learning experience. The taxonomy identifies five different levels according to the cognitive processes that the student needs to demonstrate to undertake learning. While the SOLO taxonomy describes a hierarchy of complexity in learning with each level or partial level forming a foundation on which succeeding levels are built, Biggs and Collis noted that it is possible to demonstrate different cognitive processes at different levels when learning. For example, a student may be able to demonstrate some of the processes at a multi-structural level in some subjects (e.g., describe and classify), but, in contrast, may only be able to demonstrate some of the processes at a unistructural level (e.g., perform simple instructions) in other subjects, depending on the student's perception of the learning difficulty of the subject and the instruction received. However, ultimate attainment depends on such factors as the student's intentions to learn, motivation, learning strategies, and how effective the teaching was that the student received.

FIVE LEVELS OF THE SOLO TAXONOMY

Level 1: Pre-structural. At this level, students demonstrate little understanding of the task and while they may have relevant pieces of information, they are not able to integrate them to develop logical understandings. This, in part, is due to their inability to think about or attend to more aspects of learning at once with students often not able to remember questions and answer them adequately because of the limitations of short-term memory. Students operating at the pre-structural level often fail to see logical relationships and will respond to teachers' questions with simple unelaborated detail or statements.

Level 2: Unistructural. This level is characterised by students only being able to deal with one attribute and make connections related to a task. For example, drawing the conclusion that all shapes are spheres without necessarily being able to draw conclusions about other common attributes that they may share. Students operating at this level can also be quite impulsive in the conclusions they draw, often demonstrating no need for consistency in how they think. When asked questions, students are often only able to provide one logical response, linking datum to the question.

Level 3: Multi-structural. At this level, students are able to deal with several pieces of information such as describing, classifying, and applying procedures, but they are only able to generalise in terms of a few limited and independent pieces of information at any one time. Consistent conclusions are often not evident as students consider different data independently and not in connection.

Level 4: Relational. Students may understand relationships between different data sets and how they may be integrated to give a more complete understanding of phenomena. They also demonstrate competence at being able to compare and contrast, analyse, theorise, and explain phenomena in terms of cause and effect, so they can generalise their understandings within a given context (e.g., class science lesson); however, inconsistencies may emerge when students are outside their regular context.

Level 5: Extended abstract. At this level, students are able to see relationships between and among data and generalise their observations to novel situations outside their experience. In essence, students demonstrate that they have the capacity to see situations from different perspectives, so they are able to hypothesise, critique, theorise, and transfer their understandings to new domains.

The five levels in the SOLO taxonomy are shown in Figure 6.1.

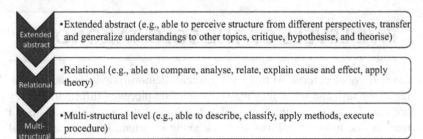

FIGURE 6.1 The five levels in the SOLO taxonomy.

INTENDED LEARNING OUTCOMES

In teaching, the focus is on teaching both the key facts and concepts, commonly called domain knowledge, and the cognitive processes or the skills and strategies that are needed to attain mastery of a subject. Domain knowledge and cognitive processes are important if students are to develop an understanding of the subject they are studying. As students move through their school years, they tend to have more exposure to different types of domain knowledge and processes so that with cognitive maturity, they develop deeper levels of understanding of the subjects they are studying. The SOLO taxonomy can be used as an important measure of structural complexity associated with learning, which can be used to provide both formative and summative assessment information on the progress students are making in different subjects.

Formative assessments occur either before or during instruction and are designed to guide teachers' planning and teaching processes. Formative assessments are not standardised or norm-referenced assessments but are based upon what students understand as a consequence of the teaching they are receiving. They are also designed to provide information on where they are experiencing difficulties and how the teaching process may need to be adjusted to overcome the difficulties that have been identified.

Many types of formative assessments can be used to provide information on students' learning, from observing students' interactions, how they engage with tasks, the types of questions they ask, and the reasons or jus-tifications they provide to explain their ideas. Formative assessments can also include curriculum-based assessments and authentic work samples that provide information on students' progress on a part of the curriculum that they have been involved in learning. These types of assessments are diagnos-tic and designed to provide both teachers and students with information on what students have accomplished and how they are managing the learning process.

Summative assessments, on the other hand, occur at the end of a period of teaching and are designed to provide teachers with information on what students have achieved. Summative assessments are not usually standardised or norm-referenced assessments although they can be. They are usually pieces of assessment that teachers have constructed to determine if students have met specific criteria. They may also include samples of students' learning, port-folios of achievement in a specific subject, exhibitions of performance (e.g., music performance, drama performance), or problem-based inquiries. In order to be able to assess the learning that has occurred from these products, teach-ers may develop rubrics to determine whether different criteria have been met, and, if so, how adequately.

Examples of the Increasing Complexity in Students' Language: Using the SOLO Taxonomy

Table 6.1 sets out the five levels of increasing complexity identified in the language Year 5 students used when they were interviewed following their experiences on an inquiry science unit that focused on investigating the three states of matter (Australian Academy of Science, 2005). The teacher used the 5Es instructional model for teaching inquiry (Bybee, 2006) that encourages students to have hands-on experiences with engaging and exploring problem issues, planning and conducting investigations, and representing and explain-ing the phenomena they had observed.

Semi-structured interviews were conducted to encourage the students to elaborate on the open-ended questions that they were asked. The emphasis in the interviews was to identify the conceptual understandings the students had developed about the different states of matter as a consequence of their par-ticipation in the science unit. Five groups of students (three per group) partici-pated in these interviews and all interviews were audiotaped (approximately 30 minutes) and fully transcribed. The transcribed interviews were then coded according to the five levels of the Structure of Observed Learning Outcomes taxonomy (Biggs & Collis, 1982). See Table 6.1 for information on the five levels and examples of students' responses to the interviewer's questions.

When the students' responses in the interviews were analysed for their level of complexity using the SOLO taxonomy, 22% were found to be at the unistructural level, 20% at the multi-structural level, 37% at the relational level, and 22% at the extended abstract level. In essence, 59% of the students' responses indicated that they understood the relationships between the different states of matter and/or were able to understand different perspectives on phenomena and could generalise their understandings to other topics.

The following are examples of how the students were able to articulate these enhanced understandings. For example, when asked what they thought of science, four students responded with the following comments:

> Cause like it helped me understand with what viscosity is and what like what liquids can be more and less viscous. (Level 4: Relational)
> Cause it's interesting to know everything that goes on in experiments and everything what the particles are doing in solids and things like that. (Level 4: Relational)
> Because it involves not just numbers but even involves um accuracy and getting to see what's going to happen and try to understand and figure out and guess what's going to be the outcome of things. (Level 5: Extended abstract)
> I liked figuring out um how the particles, like kind of worked together cause I didn't know that solid particles vibrated and the liquid particles slid over the top and I didn't know the gas particles went everywhere so I thought that was really good. (Level 5: Extended abstract)

When asked to think about the current unit and what they found interesting or intriguing about solids, liquids, and gases, three students responded as follows:

> The liquids they slip and slide all over each other and then the gases the particles bounce off each other, and in the water, you can't see, in all of them you can't see the particles are very minute. (Level 5: Extended abstract)
> Yeah, and also we had an experiment with a pen, we wrapped it in alfoil and put it in boiling hot water, to see like one that wasn't really working and then you write on the board and when you heat it it would work better because sometimes in winter the particles go closer together and then when you boil them up they go out and then they work better. (Level 5: Extended abstract)
> I, my favourite was the experiments I did particularly like the runny races because you got to see like how fast they went and like to see which one was fastest out of orange juice and that and but I also liked when we boiled we put hot the pen in hot water I like that experiment because it was one of those experiments like that you didn't know what was going to happen like with the other ones you kinda knew what was going to happen you were just finding out something [teachable moment] like if it was more viscous. I like the experiments where you're like well I wonder what's going to happen. (Level 5: Extended abstract)

Another student made the following comment about his experiences with viscosity:

> *Uh, our experiment was (the runny races) yep, there was two we did the whole class did the second one, the runny races one, we had to squirt some liquid and see which one was the most viscous, the orange juice, we had maple syrup.* (Level 4: Relational)

When asked to explain what a science diagram or representation is for, one student gave the following response:

> *Yeah, you can like draw arrows to tell you like the running races we had the chair the paper, cardboard, and the part down the bottom and we drawed in an arrow like it started and then we drawed and arrow like what we were testing.* (Level 4: Relational)

When probed to explain how they would teach someone the three states of matter, two students made the following comments:

> *I'd tell them about what the particles look like and show them like diagrams of all these things, and solids, liquids and gases.* (Level 4: Relational)
> *I'd probably explain what a particle is made up of so atoms, then what I would do is I would do explain what the particles are doing in solids, liquids and gases like they're all tight together, vibrating in a solid, in a liquid they're all sliding over each other and in a gas they're bouncing everywhere.* (Level 4: Relational)

CHAPTER SUMMARY

This chapter highlighted how the Structure of Observed Learning Outcomes taxonomy can be used to assess student reasoning, problem-solving, and learning. The SOLO taxonomy describes the increasing complexity involved in learning and includes levels that range from the pre-structural level of understanding (Level 1) where students have little knowledge or disconnected bits of information about a topic through to the extended abstract level (Level 5) that is characterised by students being able to transfer and generalise understandings to other topics and hypothesise and theorise about the reasons for different types of responses to phenomena. The focus in this chapter was on how the SOLO taxonomy was used to gauge the complexity involved in students' reasoning, problem-solving, and learning from their inquiry science activities.

In particular, the chapter drew attention to how the SOLO taxonomy could be used to identify the increasing complexity in students' language as they discussed their inquiry science experiences.

ADDITIONAL READINGS

Hooks, P. & Mills, J. (2011). *SOLO Taxonomy: A Guide for Schools Book 1.* Invercargill, NZ: Essential Resources Educational Publishers.

Hooks, P. & Mills, J. (2012). *SOLO Taxonomy: Planning for Differentiation Book 2.* Invercargill, NZ: Essential Resources Educational Publishers.

References

Adey, P. & Shayer, M. (2015). The effects of cognitive acceleration. In: L. Resnick, C. Asterhan & S. Clarke (Eds.), *Socializing Intelligence through Academic Talk and Dialogue* (pp. 127–140). Washington, DC: AERA.

Alexander, R. (2008). *Essays on Pedagogy*. London: Routledge.

Alexander, R. (2010). Dialogic teaching essentials. Retrieved from www.robinalexander.org.uk/index.php/dialogic-teaching/.

Australian Academy of Science. (2005). *Primary Connections: Linking Science with Literacy*. Canberra, Australia: Australian Academy of Science.

Bell, T., Urhahne, D., Schanze, S. & Plotezner, R. (2010). Collaborative inquiry learning: Models, tools, and challenges. *International Journal of Science Education*, *32*(3), 349–377.

Biggs, J. & Collis, K. (1982). *Evaluating the Quality of Learning: The SOLO Taxonomy* (structure of the observed learning outcomes). New York: Academic Press.

Bybee, R. (2006). Enhancing science teaching and student learning: A BSCS perspective. Proceedings of the ACER research conference: boosting science learning: What it will take. ACER research conference. *Review of Educational Research*, *64*, 1–35. Retrieved from htpp://www.acer.edu.au/research_conferences/2006.html.

Bybee, R. (2010). *The Teaching of Science: 21st-Century Perspectives*. Arlington, VA: NSTA Press.

Bybee, R. (2014). The BSCS 5 E instructional model: Personal reflections and contemporary implications. *Science and Children*, *51*(8), 10–13.

Bybee, R. (2015). *The BSCS 5 E Instructional Model: Creating Teachable Moments* (p.126). Arlington, VA: National Science Teachers' Association Press.

Carolan, J., Prain,V. & Waldrip, B. (2008). Using representations for teaching and learning science. *Teaching Science*, *54*, 18–23.

Cohen, E. (1994). Restructuring the classroom: Conditions for productive small groups. *Review of Educational Research*, *64*(1), 1–35.

Costa, A. L. & Kallick, B. (2000). Habits of mind: A developmental series. Retrieved from https://www.chsvt.org/wdp/Habits_of_Mind.pdf.

Danish, J. & Phelps, D. (2011). Representational practices by numbers: How kindergarten and first-grade students create, evaluate, and modify their science representations. *International Journal of Science Representations*, *33*, 2069–2094.

Darling-Hammond, L. & Snyder, J. (2000). Authentic assessment of teaching in context. *Teaching and Teacher Education*, *16*(5–6), 523–545.

diSessa, A. (2004). Metarepresentation: Native competence and targets for instruction. *Cognition and Instruction*, *22*(3), 293–331.

Duschl, R. & Duncan, R. (2009). Beyond the fringe: Building and evaluating scientific knowledge systems. In: S. Tobias & T. Duffy (Eds.), *Constructivist Instruction: Success of Failure?* (pp. 311–332). London: Routledge.

Facione, P. A. (1990). *Critical Thinking: A Statement of Expert Consensus for Purposes of Educational Assessment and Instruction* (The Delphi Report). Fullerton, CA: California State University.

Ford, M. J. & Forman, E. A. (2015). Uncertainty and scientific progress in classroom dialogue. In: L. B. Resnick, C. S. C. Asterhan & S. N. Clarke (Eds.), *Socializing Intelligence through Academic Talk and Dialogue* (pp.143–156). Washington, DC: AERA.

Giamellaro, M. (2014). Primary contextualization of science through immersion in content-rich settings. *International Journal of Science Education*, *36*(17), 2848–2871.

Gillies, R. (2009). *Evidence-Based Teaching: Strategies That Promote Learning* (p.193). Rotterdam, The Netherlands: Sense Publishers.

Gillies, R. (2016). *Enhancing Classroom-Based Talk: Blending Practice, Research and Theory* (p.152). London: Routledge.

Gillies, R. M. (2007). *Cooperative Learning: Integrating Theory and Practice*. Thousand Oaks, CA: Sage.

Gillies, R. & Ashman, A. (1998). Behavior and interactions of children in cooperative groups in lower and middle elementary grades. *Journal of Educational Psychology*, *90*(4), 746–757.

Gillies, R. & Baffour, B. (2017). The effects of teacher-introduced multimodal representations and discourse on students' task engagement and scientific language during cooperative, inquiry-based science. *Instructional Science*, *45*(4), 493–513.

Gillies, R. & Boyle, M. (2006). Ten Australian elementary teachers' discourse and reported pedagogical practices during cooperative learning. *The Elementary Journal*, *106*(5), 429–451.

Gillies, R. & Khan, A. (2008). The effects of teacher discourse on students' discourse, problem-solving and reasoning during cooperative learning. *International Journal of Educational Research*, *47*(6), 323–340.

Gillies, R. & Khan, A. (2009). Promoting reasoned argumentation, problem-solving and learning during small-group work. *Cambridge Journal of Education*, *39*(1), 7–27.

Gillies, R., Nichols, K. & Burgh, G. (2011). Promoting problem-solving and reasoning during cooperative inquiry science. *Teaching Education*, *22*(4), 429–455.

Herreid, Clyde Freeman (1994). Case studies in science – A novel method of science education. *Journal of College Science Teaching*, 221–229. Retrieved from http://sciencecases.lib.buffalo.edu/cs/training/.

Harris, C. & Rooks, D. (2010). Managing inquiry-based science: Challenges in enacting complex science instruction in elementary and middle school classrooms. *Journal of Science Teacher Education*, *21*(2), 227–240.

Herrenkohl, L., Tasker, T. & White, B. (2011). Pedagogical practices to support classroom cultures of scientific inquiry. *Cognition and Instruction*, *29*(1), 1–44.

Howe, C. & Abedin, M. (2013). Classroom dialogue: A systematic review across four decades of research. *Cambridge Journal of Education*, *43*(3), 325–356.

Hubber, P., Tytler, R. & Haslam, F. (2010). Teaching and learning about force with a representational focus: Pedagogy and teacher change. *Research in Science Education*, *40*(1), 5–28.

Huff, K. & Bybee, R. (2013). The practice of critical discourse in science classrooms. *Science Scope, 36*(9), 30–34.

Johnson, D. & Johnson, F. (2009). *Joining Together: Group Theory and Group Skills* (10th ed.). Boston, MA: Allyn and Bacon.

Johnson, D. & Johnson, R. (2002). Learning together and alone: Overview and meta-analysis. *Asia Pacific Journal of Education, 22*(1), 95–105.

Johnson, D., Johnson, R. & Houlbec, E. (2009). *Circles of Learning* (6th ed.). Edina, MN: Interaction Book Company.

Kind, P. & Osborne, J. (2017). Styles of scientific reasoning: A cultural rationale for science education? *Science Education, 101*(1), 8–31.

King, A. (1997). Ask to think-tel why: A model of transactive peer tutoring for scaffolding higher level complex learning. *Educational Psychologist, 32*(4), 221–235.

King, A. (1999). Discourse patterns for mediating peer learning. In: A. M. O'Donnell & A. King (Eds.), *Cognitive Perspectives on Peer Learning*. Mahwah, NJ: Lawrence Erlbaum Publishers.

Klein, P. & Kirkpatrick, L. (2010). Multimodal literacies in science: Currency, coherence and focus. *Research in Science Education, 40*(1), 87–92.

Krajcik, J. & Sutherland, L. (2010). Supporting students in developing literacy in science. *Science, 328*(5977), 456–459.

Lee, O., Hart, J., Cuevas, P. & Enders, C. (2004). Professional development in inquiry-based science for elementary teachers of diverse student groups. *Journal of Research in Science Teaching, 41*(10), 1021–1043.

Lemke, J. (2004). The literacies of science. Retrieved from http://jaylemke.squarespa ce.com/storage/Literacies-of-science-2004.pdf.

Lin, T., Hsu, Y., Lin, S., Changlai, M., Yang, K. & Lai, T. (2012). A review of empirical evidence on scaffolding for science education. *International Journal of Science and Mathematics Education, 10*(2), 437–455.

Lipman, M. (1988). *Philosophy Goes to School*. Philadelphia, PA: Temple University Press.

Llewellyn, D. (2014). *Inquire within: Implementing Inquiry and Argument-Based Science Standards in grades 3–8*. Thousand Oaks, CA: Corwin.

Lou, Y., Abrami, P., Spence, J., Poulsen, C., Chambers, B. & d'Apollonia, S. (1996). Within-class grouping: A meta-analysis. *Review of Educational Research, 66*(4), 423–458.

Lucariello, J., Nastasi, B., Anderman, E., Dwyer, C., Ormiston, H. & Skiba, R. (2016). Science supports education: The behavioural Research Base of Psychology's top 20 principles for enhancing teaching and land learning. *Mind, Brain, and Education, 10*(1), 55–67.

Mayer, R. (2002). Cognitive theory and the design of multimedia instruction: An example of the two-way street between cognition and instruction. *New Directions in Teaching and Learning, 89*, 55–71.

Mercer, N. (2008). Talk and the development of reasoning and understanding. *Human Development, 51*(1), 90–100.

Mercer, N. & Dawes, L. (2014). The study of talk between teachers and students, from the 1970s until the 2010s. *Oxford Review of Education, 40*(4), 430–455.

Mercer, N. & Littleton, K. (2007). *Dialogue and the Development of Children's Thinking: A Sociocultural Approach*. London: Routledge.

Mercer, N. & Sams, C. (2006). Teaching children how to use language to solve maths problems. *Language and Education*, *20*(6), 507–528.

Mercer, N., Wegerif, R. & Dawes, L. (1999). Children's talk and the development of reasoning in the classroom. *British Educational Research Journal*, *25*(1), 95–111.

Metz, K. (2008). Narrowing the gulf between the practices of science and the elementary science classroom. *The Elementary School Journal*, *109*(2), 138–161.

National Research Council. (2012). *A Framework for K-12 Science Education: Practices, Cross-Cutting Concepts, and Core Ideas*. Washington, DC: The National Academies Press.

National Science Teachers Association. (2004). NSTA position statement: Scientific inquiry. Retrieved from http://www.nsta.org/about/positions/inquiry.aspx/.

Newman, W., Abell, S., Hubbard, P., McDonald, J., Ottaala, J. & Martini, M. (2004). Dilemmas of teaching inquiry in elementary science methods. *Journal of Science Teacher Education*, *15*(4), 257–279.

Osborne, J. (2006). Towards a science education for all: The role of ideas, evidence and argument. *Boosting Science Learning: What It Will Take. ACER Research Conference*. Retrieved from http://www.acer.edu.au/research_conferences/2006.html.

Pearson, P. D., Moje, E. & Greenleaf, C. (2010). Literacy and science: Each in the service of the other. *Science*, *328*(5977), 459–463.

Piaget, J. (1950). *The Psychology of Intelligence*. London: Routledge & Kegan.

Pouw, W., van Gog, T. & Paas, F. (2014). An embedded and embodied cognition review of instructional manipulatives. *Educational Psychology Review*, *26*(1), 51–72.

Rennie, L. (2005). Science awareness and scientific literacy. *Teaching Science*, *51*(1), 10–14.

Resnick, L., Michaels, S. & O'Connor, C. (2010). How (well structured) talk builds the mind. In: D. Pressis & R. Sternberg (Eds.), *Innovations in Educational Psychology: Perspectives on Learning, Teaching and Human Development*. New York: Springer.

Reznitsakaya, A., Anderson, R. & Kou, L. (2007). Teaching and learning argumentation. *The Elementary School Journal*, *107*(5), 449–472.

Reznitskaya, A., Glina, M., Carolan, B., Michaud, O., Rogers, J. & Sequeira, L. (2012). Examining transfer effects from dialogic discussions to new tasks and contexts. *Contemporary Educational Psychology*, *37*(4), 288–306.

Rojas-Drummond, S., Perez, V., Velez, M., Gomez, L. & Mendoza, A. (2003). Talking for reasoning among Mexican primary school children. *Learning and Instruction*, *13*(6), 653–670.

Topping, K. & Trickey, R. (2014). The role of dialogue in philosophy for children. *International Journal of Educational Research*, *63*, 69–78.

Topping, K., Trickey, S. & Cleghorn, P. (2019). *A Teacher's Guide to Philosophy for Children*. New York: Routledge (p. 175).

Tytler, R. (2007). Re-imagining science education: Engaging the students in science for Australia's future. *Australian Education Review*. Camberwell, Vic: ACER.

Veermans, M., Lallimo, J. & Hakkaraienen, K. (2005). Patterns of guidance in inquiry learning. *Journal of Interactive Learning Research*, *16*, 179–194.

Zuckerman, G., Chudinova, E. & Khavkin, E. (1998). Inquiry as a pivotal element of knowledge acquisition within the Vygotskian paradigm: Building a science curriculum for the elementary school. *Cognition and Instruction*, *16*(2), 201–233.

Index

A

Ask to Think-Tel Why questioning approach
  hint questions, 50
  metacognitive questions, 51
  probing questions, 50
  review questions, 50
  thought-provoking questions, 51
Assessments, 94, 96
  formative, 95, 96, 105
  summative, 95–97, 105

C

Case-based learning (CBL), 94–96
Classroom-based discourse, 5
Cognitive flexibility, 76
Cognitive processes, 51, 100, 102, 104
Cooperation groups, strategies for
      constructing, 93–94
Cooperative group work, visual
      organiser, 19
Cooperative learning, 82–83
  benefits of, 83–85
  group processing
    promoting, 92–93
    and reflecting, 92
  groups, 84–85
  key elements in, 86–93
  positive interdependence, 86–87
  promotive interaction, 87–88
  research on assessing, 96
  strategies for assessing, 94–96
Cooperative learning activities, 11–14
  structuring, 19

D

Deeper conceptual understandings, 40–42
Dialogic classrooms, 65–66
Dialogic interactions, 62, 64, 79; *see also*
    Dialogic teaching

classroom discussions, expectations, 71
in cooperative group setting, 68–70
evidence-based explanations or solutions,
    opposing, 72
feedback, 73–74
immature conceptions, 72–73
interpersonal relationships and
    communication, 74
learning objectives, 71–72
learning situations, 74
observe and explain, opportunities, 72
practice learning, 73
practices of science, students, 70–71
prior knowledge with evidence,
    connecting, 73
scientific argumentation, 75–76
strategies to promote, 70–76
teach critical discourse, 74–75
Dialogic strategies
  critical thinking skills, 77–78
  for students, 76–78
Dialogic teaching, 23, 63–70, 77
  collective, 65
  cumulative, 65
  example of, 67–68
  purposeful, 65
  reciprocal, 65
  supportive, 65
Discipline-specific language, 2
Discourse-intensive instruction, 63
Domain knowledge, 104

E

Earthquakes, 46–48
  team word web, 17
Elaboration phase, 8–9
Embodied representations, 22, 40
Engagement phase, 6–7
Evaluation phase, 9–10
Exploration phase, 7
Extended abstract level, 40–41

F

5Es instructional model, 6, 24, 26, 40
Formal cooperative learning, 85
Formative assessments, 95, 96, 105
Friendship groups, 94, 96

G

Guessing game, 39

H

Higher-level thinking, 10, 20
Hybrid texts, 44

I

Informal cooperative learning, 85
Initiation-response feedback (I-R-F), 5
Inquiry-based science, 2–5
    background, 1–2
    challenges implementing, 18–19
    to challenge thinking, 5–10
    complex tasks characteristics, 16–17
    group evaluation, 16
    group's action plan, 16
    individual reflection activity, 15
    strategies promoting, 11–17
        brainstorm with peer, 11
        complex tasks, 14
        cooperative learning activities, 11–14
        group composition, 13–14
        group size, 13
        listen and recall, 12
        paired activity, 11–12
        paired-questioning, 12–13
        simple tasks, 14
        think-pair-share, 13
        2-minute review, 12
    students' learning evaluation, 16–17
Inquiry-based science lessons, 26–40
    elaborate, 36–37
    engage, 27–30
    evaluate, 37–40
    explain, 33–36
    explore, 30–33
Inquiry learning, 3, 40
Inquiry process, 1, 3
    steps in, 4
Intended learning outcomes, 104–107
I-R-F, *see* Initiation-response feedback

K

Know, Learned, and Questions raised (KLQ)
    chart, 10

L

Language representations, 25–26, 40
    basic language, 25
    encouragers, 25
    maintenance language, 26
    mediation, 26
"Learning talk repertoire," 76
Linguistic tools, 56–58, 62, 70

M

Method
    context for study, 24
    data collection, 25
    inquiry-based science unit, 24–25
    teacher measures
        embodied representations, 25
        language representations, 25–26
        student interviews, 26
        visual representations, 25
Mixed-ability groups, 93
Multimodal representations, 22

N

Narrative, 48

P

Philosophy for Children (P4C), 60–61
Piagetian Theory of Cognitive
    Development, 102

R

Relational level, 40
Representational competence, 22
Representations
    case study, purpose, 24
    types, 22–24

S

Science-as-practice perspective, 3
Science education goals, 2
Science education programme elements, 2

Scientific community, 5
Scientific discourse, promoting, 63–79
Scientific explanation phase, 7–8
Scientific ideas, 3
Scientific inquiry, 3
Scientific literacy, 44–61
  Accountable Talk, 58–59
  anchoring learning, 46–48
  audience participation, encouraging,
      54–56
  background, 43–44
  children's understandings, challenging
      questions, 50–51
  developing, 43–62
  earthquakes, measuring, 45
  Exploratory Talk, 59–60
  knowledge and experience, new ideas,
      44–45
  linguistic tools, student discussion,
      56–58
  literacy principles, 44–50
  multiple representations, connecting,
      48–49
  Philosophy for Children (P4C), 60–61
  science discourse and, 53–54
    students' engagement, 49–50
  science ideas, opportunities, 49

  stems and cognitive processes,
      question, 51–53
Semi-structured interviews, 105
Small, cooperative group instruction, 84
Socratic Questioning, 77
Soil liquefaction, 47–48
Structure of Observed Learning Outcomes
      (SOLO) taxonomy, 26, 40, 99,
      100–102, 107, 108
  five levels of, 103–104
  intended learning outcomes, 104–107
  students' language, complexity, 105–107
Student discourse, fostering, 23
Summative assessments, 95–97, 105

T

Think, Want, Learnt, How (TWLH) chart,
      6–7

V

Visual representations, 40

W

"What's the Matter" unit, 24, 27, 42